John George Wood

Strange Dwellings

Being a Description of the Habitations of Animals

Elibron Classics
www.elibron.com

Elibron Classics series.

ISBN 0-543-85180-X (paperback)
ISBN 0-543-85179-6 (hardcover)

This Elibron Classics Replica Edition is an unabridged facsimile
of the edition published in 1873 by Longmans, Green, and Co., London.

STRANGE DWELLINGS

LONDON: PRINTED BY
SPOTTISWOODE AND CO., NEW-STREET SQUARE
AND PARLIAMENT STREET

NESTS OF BAYA SPARROW (*Ploceus baya*). See page 148.

STRANGE DWELLINGS

BEING A DESCRIPTION

OF

THE HABITATIONS OF ANIMALS

ABRIDGED FROM 'HOMES WITHOUT HANDS'

BY THE

REV. J. G. WOOD, M.A. F.L.S. &c.

TRAP-DOOR SPIDER AND NEST

WITH DESIGNS
by
W. F. KEYL, J. B. ZWECKER, AND E. SMITH

NEW EDITION

LONDON
LONGMANS, GREEN, AND CO.
1873

PREFACE.

THE OBJECT of this work is so fully given in the Title-page, that little more remains to be said in the Preface.

Beginning with the simplest and most natural form of habitation, namely, a burrow in the ground, the work proceeds in the following order:—2nd, those creatures that suspend their homes in the air; 3rd, those that are real builders, forming their domiciles of mud, stones, sticks, and similar materials; 4th, those which make their habitations beneath the surface of the water, whether salt or fresh; 5th, those that live socially in communities; 6th, those which are parasitic upon animals or plants; 7th, those which build on branches. The last chapter treats of miscellanea, or those habitations which could not be well classed in either of the preceding groups.

In all these classes a definite order has been preserved, the Mammalia having precedence, and being followed in regular order by the other members of the group. Thus, in the first few chapters, which treat of the Burrowers, the following system has been observed: First comes Man, the chief of all the mammalia, and in due zoological order follow the Moles and Shrews, the Foxes, the Weasels, the Rodents, and the Edentates. The White Bear alone is removed from its legitimate place, on account of its singular habitation in the snow. The Burrowing Birds come next in order, those which burrow in the earth taking precedence of those which make holes in wood. Burrowing Reptiles follow next in order; and then come the Burrowing Invertebrates, headed by the Crustacea. The same system is followed throughout, so as to give the reader a clear and definite idea of the subject.

On perusing the work, the attentive reader will probably discover that various animals are placed in one class when they might very well be in another. The reason is, that many creatures, such as the wasp, the ant, the squirrel, &c., might with equal propriety find a place in several of these classes, and I have therefore placed them in that class of which some peculiarity in nest-making renders them fit illustrators.

I must now return my thanks to the many friends who have assisted me in the work, by the loan or gift of specimens, or by affording valuable information. Among them I must especially mention J. GOULD, Esq., who kindly took an interest in the ornithological portion of the work; F. SMITH, Esq., of the British Museum; and the late CHARLES WATERTON, Esq., who permitted me the use of his museum, and gave me much interesting and useful information.

CONTENTS.

CHAPTER		PAGE
I.	Burrowing Mammalia	1
II.	Burrowing Birds	32
III.	Burrowing Reptiles	44
IV.	Burrowing Invertebrates	47
V.	Burrowing Molluscs	53
VI.	Burrowing Spiders	64
VII.	Burrowing Insects (*Hymenoptera*)	73
VIII.	Burrowing Beetles	91
IX.	Wood-boring Insects	102
X.	Pensile Mammalia	117
XI.	Pensile Birds	121
XII.	Pensile Birds—*continued*	127
XIII.	Pensile Birds—*continued*	137
XIV.	Pensile Insects	151
XV.	Builders	180
XVI.	Building Birds	184

CHAPTER		PAGE
XVII.	Building Birds—*continued*	193
XVIII.	Building Insects	207
XIX.	Sub-Aquatic Nests (*Vertebrates*)	225
XX.	Sub-Aquatic Nests (*Invertebrates*)	228
XXI.	Social Habitations	245
XXII.	Social Birds	251
XXIII.	Social Insects	256
XXIV.	Social Insects—*continued*	273
XXV.	Parasitic Nests	287
XXVI.	Parasitic Nests—*concluded*	309
XXVII.	Branch-Building Mammalia	320
XXVIII.	Feathered Branch-Builders	323
XXIX.	Feathered Branch-Builders—*concluded*	344
XXX.	Branch-Builders (*Spiders and Insects*)	360
XXXI.	Miscellanea	379
INDEX		407

ILLUSTRATIONS.

	PAGE
Nests of Baya Sparrow (*Ploceus baya*)	*Frontispiece*
Trap-door Spider and Nest	*Titlepage*
Fortress of the Mole	2
The Rabbit Warren	17
Canada Pouched Rat (*Plan of Burrow*)	20
Polar Bear	21
Pichiciago	23
Aard Vark	26
Sand Martins	33
Puffin	38
Pholas in Wood—Lithodomus—Razor Shell—Holes of Pholas in Rock	58
Shipworm	61
Nest of Trap-door Spider (from a Specimen in the collection of Lieut.-Col. C. J. Cox)	69
Myrmeleon—Ampulex—Scolia	80
Bombus Terrestris—Bombus Lapidarius	83
Cocoons of Scarabæus and Goliath	95
Scolytus	103
The Tailor Birds	123
Ptilotus Sonorus—Entomophila Picta—Entomophila Albogularis—Sericornis Citreogularis—Origma Rubricata	128
Swallow Dicæum	134
Lanceolate Honey-eater (*Plectorhynchus lanceolatus*) . .	136
Little Hermit (*Phaethornis eremita*)	138
Sawbill Humming Bird—Brazilian Wood Nymph—White-sided Hill Star	140
Crested Cassique—Baltimore Oriole	143
Tatua Morio	152
Nectarinia—Myrapetra	155

PAGE
Nests of Polistes 163
Oiketicus and Atlas Moth 165
Leaf Miners and Rollers 175
Oven Bird 185
Fairy Martin—Pied Grallina 187
The Long-tailed Titmouse 196
The Bower Bird 203
Trypoxylon and Pelopæus 214
Water Spider 230
Caddis 234
Sociable Weaver Bird 252
Polybia 257
Cells of the Honey Bee 262
Hornet 268
Driver Ants 275
Mud Wasp 285
Parasitic Insects. Cocoon of Oak-egger Moth (*Cryptus fumipennis.*)—Cocoon of Puss Moth (*Paniscus glaucopterus*)—Cocoon from New South Wales (*Pimpla*)—Cocoon of Goat Moth (*Lamprosa setosa*) 294
British Galls. Leaf Galls of Oak—Bedeguar of Rose—Galls of Cynips Kollari—Cynips Kollari (*slightly magnified*)—Currant Galls of Oak—Oak Apple . 298
Nest of Goldfinch 332
Golden Orioles and Nest 336
Ringdove and Nest 340
Water Hen and Nest 342
Fiery Topaz and Hermit 351
Pensile Spider's Nest 361
Mischocyttarus—Icaria—Raphigaster 363
Apoica 370
Raft Spider 380
Mouse-Nest in Bottle 389
Termite Cell 391
Edible Swallow 400
Eagle 402

STRANGE DWELLINGS.

CHAPTER I.

BURROWING MAMMALIA.

Introduction—MAN as a Burrower—The MOLE and its Dwelling—Difficulty of observing its Habits—Complicated structure of its Fortress, and its Uses—Character of the Mole—Adaptation of its Form to its mode of Life—Common Objects—The SHREW MOLE, ELEPHANT SHREW, and MUSK RAT—The ARCTIC FOX—Structure of its Limbs—Form of its Burrow—The Common FOX—Mode of Burrowing and economy of Labour—The young Family—The WEASELS—The BADGER and its Burrow—The PRAIRIE DOG, or WISH-TON-WISH—Dog-towns—Unpleasant Intruders—The RABBIT, and the Warren—The CHIPPING SQUIRREL—Curious form of its Dwelling—The POUCHED RAT—The WHITE BEAR—Its curious Dwelling—Snow as a Shelter—The PICHICIAGO—Its Form, Armour, and Burrow—The MANIS—The AARD VARK, Its Food and Dwelling—The MALLANGONG—Its strange Habits and its Burrow—The PORCUPINE ANT-EATER—Its burrowing Powers.

AT some period of their existence, many of the higher animals require a Home, either as a shelter from the weather, or a defence against their enemies. Of all forms of habitation, the simplest is a burrow, whether beneath the surface of the ground, or into stone, wood, or any other substance.

The lowest grades of human beings are found to adopt this easy and simple substitute for a home, and the Bosjesman of the Cape, and the 'Digger' Indian of America, alike resort to so obvious an expedient.

Human habitations, however, do not come within the scope of the present work, which is restricted to those homes that are constructed without the aid of hands, and are planned, not by reason, but by instinct. We pass, therefore, from the handiwork of man to those dwellings which are constructed with feet or jaws or beaks, and which are never marred by incompetence or improved by practice.

FORTRESS OF THE MOLE.

Of all the mammalia, the Mole is entitled to take the first place in our list of burrowers.

This extraordinary animal does not merely dig tunnels in the ground and sit at the end of them, but forms a complicated subterranean dwelling-place, with chambers, passages, and other arrangements of wonderful completeness. It has regular roads leading to its feeding-grounds; establishes a system of communication as elaborate as that of a modern railway, or to be more correct, as that of the subterranean network of metropolitan sewers; and is an animal of varied accomplishments.

It can run tolerably fast, it can fight like a bulldog, it can capture prey under or above ground, it can swim fearlessly, and it can sink wells for the purpose of quenching its thirst. It is, indeed, a most interesting animal, and our comparatively small knowledge of its habits gives promise of much that is yet to be made known.

Take the Mole out of its proper sphere, and it is as awkward and clumsy as the sloth when placed on level ground, or the seal when brought ashore. Replace it in the familiar earth, and it becomes a different being,—full of life and energy, and actuated by a fiery activity which seems quite inconsistent with its dull aspect and seemingly inert form. The absence of any external indication of eyes communicates a peculiar dulness to the creature's look, and the peculiar formation of the fore limbs gives an indescribable awkwardness to its gait.

I have always taken much interest in this animal, and have watched many of its habits, as far as can be done under the very untoward circumstances that always must exist when the animal to be watched is essentially subterranean in its habits. The Mole cannot develop its nature unless it is buried below the surface of the ground, and when it is there, we cannot see it. Many marine and aquatic animals can be tolerably watched by placing them in the aquarium; but when they take to burrowing, they put an effectual stop to investigation.

WE all know that the Mole burrows under the ground, and that it raises those little hillocks with which we are so familiar, but we do not generally know the extent or variety of its tunnels, or that the animal works upon a regular system, and does not burrow here and there at random. How it manages to form its burrows in such admirably straight lines is not an easy problem, because it is always in black darkness, and we know of nothing which can act as a guide to the animal. As for ourselves and other eye-possessing animals, to walk in a straight line with closed eyelids is almost an impossibility, and every swimmer knows the difficulty of keeping a straight course under water, even with the use of his eyes.

The ordinary mole-hills, which are so plentiful in our fields, present nothing particularly worthy of notice. They are the shafts through which the quadrupedal miner ejects the materials which it has scooped out, as it drives its many tunnels through the soil, and if they be carefully opened after the rain has consolidated the heap of loose material, nothing more will be discovered than a simple hole leading into the tunnel. But let us strike into one of the large tunnels, as any mole-catcher will teach us, and follow it up until we come to the real abode of the animal.

A section of this extraordinary habitation is given in the illustration. The hill under which this domicile is hidden is of considerable size, but is not very conspicuous, because it is always placed under the shelter of a tree, a shrub, or a suitable bank, and would not be discovered but by a practised eye. The subterraneous abode within the hillock is so remarkable that it involuntarily reminds the observer of the well-known maze, with which the earliest years of youth have been puzzled throughout many successive generations.

The central apartment, or keep, if we so term it, is a nearly spherical chamber, the roof of which is nearly on a level with the earth around the hill, and therefore situated at a considerable depth from the apex of the heap. Around this keep are driven two circular passages, or galleries, one just level with the ceiling and the other at some height above. The upper circle is much smaller than the lower. Five short descending passages connect the galleries with each other, but the only entrance into the keep is from the upper gallery, out of which three passages lead into the ceiling of the keep. It will be seen, therefore, that when a Mole enters the house from one of his tunnels, he has first to get into the lower gallery, to ascend thence to the upper gallery, and so descend into the keep.

There is, however, another entrance into the keep from below. A passage dips downwards from the centre of the chamber, and then, taking a curve upwards, opens into one of the larger tunnels, or high roads, as they have been appropriately termed. It is a noteworthy fact, that the high roads, of which there are seven or eight, radiating in different directions, never open into the gallery opposite one of the entrances into the upper gallery. The Mole, therefore, is obliged to turn to the right or left as soon as it enters the domicile, before it can find a passage to the upper gallery.

By continual pressure of the Mole's fur, the walls of the passages and the roof of the central chamber become quite smooth, hard, and polished, so that the earth will not fall in even after the severest storm.

Wonderful as is this subterranean habitation, it is not the only one which is constructed by the animal. It may be well adapted to a solitary individual, but it is not at all suited for a family, for whom a more extended nursery must be provided. The nursery is much simpler than the habitation, consisting merely of a large chamber, in which is laid a considerable mass of dried grass, the young blades of corn being sometimes employed for that purpose. The Mole chooses for this purpose the spot where two or more passages intersect each other, so that in case of alarm, the mother and young may escape in the direction which seems farthest removed from danger. This nursery is almost invariably placed at some distance from the fortress.

About the middle of June, or commencement of July, the Moles begin to fall in love, and are as furious in their attachments as in all other phases of their nature. At that time, two male Moles cannot meet without a mortal jealousy, and they straightway begin to fight, scratching, tearing, and biting with such insane fury, that they seem to be unconscious of everything but the heat of battle. Not content with fighting in their burrows, they often emerge into the open air, and may then be caught without the least difficulty. A few days before writing this account, I heard that a pair of Moles were thus taken in the fields near Erith, and one of my friends made a similar capture on Shooter's Hill.

Indeed, the whole life of the Mole is one of fury, and he eats like a starving tiger, tearing and rending his prey with claws and teeth, and crunching audibly the body of the worm between the sharp points. Some writers say that the Mole eats snails and other molluscs, but I am disposed to doubt that assertion. I have kept several Moles and never saw them eat anything but worms. They even rejected the julus millipede, kicking it aside with utter contempt.

It is also asserted that the Mole skins the worm before he eats it, 'stripping the skin from end to end, and squeezing out the contents of the body.' To prove a negative is proverbially a difficult task, and therefore I will not venture to say that the Mole does not trouble himself about stripping off the skin of the worm. I do not see how he could do so, for even with the assistance of knives, scissors and forceps, such a task presents many difficulties, and how the Mole is to succeed in such an undertaking with no tools but his teeth and claws, I cannot comprehend. No Mole that I have ever seen, gave the slightest indication of skinning or emptying the worm, but proceeded without the least ceremony to devour the writhing prey, and then looked out for another victim.

It is hardly possible to conceive, and quite impossible to describe the fury with which the Mole eats. It hunches its back in a most curious manner, retracts the head between the shoulders, and uses its fore paws to assist it in pushing the worm into its jaws. In this respect there is a singular resemblance between the Mole and the carnivorous chelodines of America. I have kept several of them, and have always noticed that they ate exactly after the fashion employed by the Mole, seizing their food in their jaws, and tearing it to pieces by the aid of the armed fore paws—one foot being applied at each side of the mouth, so as to push the food forwards, while the head draws it back.

How the Mole assumes this peculiar attitude I cannot conceive. I have often seen it engaged in eating, and have sketched the creature while so employed; but, when the Mole has been dead, I have been unable to place it in the proper attitude, though anxious to do so in order that the artist might be able to make his drawing properly.

From seeing the animal eat, I can readily conceive the fury with which it must be animated when it fights, and can perfectly appreciate the truth of the assertion, that it has been observed to fling itself upon a small bird, to tear its body open, and to devour it while still palpitating with life.

Nothing short of this fiery energy could sustain an animal in the lifelong task of forcing itself through the solid earth; and it may well be imagined that when two male Moles of equal strength happen to meet, the combat must be of the most furious kind.

To those who are accustomed only to look at animals from their own stand-point, these battles may appear too insignificant to attract attention; but to the eye of a naturalist, who instinctively identifies himself with the nature of the animals which he is observing, these combats lose all their insignificance, and even partake in some degree of the sublime. Size is only of relative importance; and, in point of fact, a battle between two Moles is as tremendous as one between two lions, if not more so, because the Mole is more courageous than the lion, and, relatively speaking, is far more powerful and armed with weapons more destructive.

On looking over the list of burrowing mammalia, the observer cannot but be struck with the wonderful manner in which they emerge from the earth with unsoiled fur. This capability is the more remarkable in the animal now under consideration, because it is continually engaged in making new tunnels, and is not content merely to pass up and down a passage already excavated. The sides of the passages, which are popularly known as the high roads, are by degrees worn quite smooth by the attrition of the Mole's body, so that in them there is little danger of injury accruing to the fur. But that an animal should be able to pass unsoiled through earth of all textures is a really remarkable phenomenon, which is partly to be explained by the character of the hair, and partly by that of the skin.

The hair of the Mole is notable for its velvety aspect, and its want of 'set.' The tips of the hairs do not point in any particular direction, but may be pressed equally forwards or backwards or to either side. The microscope reveals the cause of this peculiarity. The hair is extremely fine at its exit from the skin, and gradually increases in thickness. When it has reached its full width, it again diminishes. This alternation of tenuity and thickness occurs several times in each hair, and gives the peculiar velvet-like texture with which we are all so familiar. There is scarcely any colouring matter in the slender portions of the hair, and the characteristic changeability of the blackish-brown hues is owing to this structure.

Perhaps the reader may not have noticed that when the fur of the Mole has been thoroughly cleansed, it has a strong iridescence in certain lights, assuming various beautiful tints, among which a ruddy copper is the most prevalent.

Another reason for the cleanliness of the fur is the strong, though membranous muscle beneath the skin. While the Mole is engaged in tunnelling, particularly in loose earth, the soil falls upon the fur, and for a time clings to it. But, at tolerably regular intervals, the creature gives the skin a sharp and powerful shake, which throws off at once the whole of the mould that has collected upon the fur. Some amount of dust still retains its hold, for, however clean the fur of a Mole may seem to be, if the creature be placed for an hour in water, a considerable quantity of earth will be dissolved away, and fall to the bottom of the vessel. The improvement in the fur after being well washed with soft tepid water and soap, is almost incredible.

I have given much space to the Mole on account of its many claims to our notice. Had the creature been a rare and costly inhabitant of the tropics, how deep would have been the interest which it excited. How the scientific world would have crowded to see the marvellous structure of a skeleton wherein are several accessory bones, and which exhibits peculiarities hitherto found only in fossil remains. How great would have been the admiration evoked by its soft, velvet-like fur, its tiny eyes deeply hidden in the fur, so as to be sheltered from the earth through which the animal is continually making its way, the strange mixture of strength and softness in the palms of its fore feet, and the elastic springiness of its nose.

But, because it is a native of our own country, and to be found in every field, there are but few who care to examine a creature so common, or who experience any feelings save those of contempt or disgust, when they see a Mole making its way over the ground in search of a soft spot in which to burrow, or pass by the place where the mole-catcher has strung up his victims on the trees as Louis XI. was accustomed to suspend the bodies of those who had committed the crime of trespassing on the royal domains. For my own part, I am but too glad that such wonderful beings *are* common, and am thankful for so many opportunities of studying the works of Him who has made the lowly Mole as carefully as the lordly man.

THERE are many other burrowing animals allied to the mole; and although it will be impossible to give illustrations of their burrows, they ought not to be passed by without a casual notice.

The Shrews, for example, are among the burrowers, and although their eyes are full and round, their fore quarters of ordinary proportions, and their fore feet of the usual shape, there is something about the head, with its long mobile snout, which strongly reminds the observer of the same member in the mole. These pretty little creatures reside within their burrows during the day, and are therefore seldom seen in a living state, except by those who are in the habit of traversing the country by night in search of specimens. Dead Shrews are common enough, having probably been killed by predatory animals, but left uneaten on the ground, in consequence of the powerful odour which they evolve.

At the end of the burrow the Shrew makes its nest, which is composed of dry grasses and other herbage, and is of a partly globular form.

The SHREW MOLE of North America (*Scalops aquaticus*), is one of the best burrowers among this family, scarcely yielding to the mole itself in the extent of the tunnels which it excavates. Like the mole, it drives its burrows below the surface of the ground, throws up hillocks at intervals, and feeds chiefly on earthworms. The eyes of this creature are very minute, and deeply hidden in the soft fur. Unlike the mole, however, it is in the constant habit of coming to the surface of the ground, and passing into the full blaze of the noontide sun. At that time of day the animal may be caught by driving a spade under it, so as to cut off its retreat, and by flinging it to some distance from its tunnel.

Mr. Peale mentions that a Shrew Mole in his possession was able to bend the snout to such an extent as to force food into its mouth. The European mole, flexible as is its mobile snout, possesses no such power, but is obliged to perform that task with its fore paws.

Then, there is the ELEPHANT SHREW of Southern Africa (*Macroscelides typicus*), a thick-furred, long-snouted, short-eared burrower, which has a rather remarkable method of sinking its tunnels, first boring a nearly perpendicular shaft, and then driving its burrow at an angle. It is not so devoted to a subterranean existence as either of the preceding animals, and loves to come out of its burrow and bask in the genial sunbeams. It is, however, as wary as the rest of its kindred, and at the least alarm darts off to its subterranean fastnesses. While basking in the warm rays, it generally sits erect, facing the sun, so as to receive every ray.

Our last example of the Shrews is the remarkable animal which is popularly called the MUSK RAT (*Myogalea moschata*), though it is an insectivorous animal, and far removed from the rodents. The river Wolga is the favourite resort of this curious quadruped, which seems to hate dry land as much as the beaver, and to spend the greater part of its time in the water. The Musk Rat is an admirable burrower, making its tunnels of considerable length, some of them extending to a distance of twenty feet. There is only one entrance, which is always below the water; and the burrow rises gradually upwards, so that at the extremity the animal is lodged on dry ground. It is instinctively careful to avoid too close a proximity to the surface of the earth, lest the roof of its home might fall, and disclose the interior to the unwelcome light.

THE FOX is a well-known burrower, its 'earth' being familiar to many by sight, and to all by name.

Few persons, who do not know the history of the Fox, would believe it to be capable of forming excavations of such extent. The fore feet of the mole are clearly formed for digging, their sharp claws penetrating the earth, their broad palms acting as shovels, and their powerful muscles giving the needful force. These limbs are essentially used for digging, and are but little employed as means of locomotion. But the Fox is an admirable runner, as any hunter can avouch, and its fore limbs are formed for speed and endurance, their length enduing them with the one quality, and their muscular lightness with the other. Yet, just as the digging limbs of the mole are used for locomotion, and enable the animal to proceed at no contemptible speed; so the running limbs of the Fox are used for digging, and enable the creature to excavate burrows of no contemptible dimensions.

The ARCTIC FOX (*Vulpes lagopus*), an animal which dwells in the polar regions, is notable for the extent and structure of the burrow. In order to shield itself from the inclemency of the climate, it digs to a considerable depth; and it is rather remarkable that a solitary burrow is seldom found, twenty or thirty Foxes generally sinking their tunnels in close proximity to each other.

Perhaps this semi-sociality may be accounted for in a very simple manner, namely, the suitability of some particular piece of ground, to which the Foxes flock by instinct, and in which they drive as many burrows as the ground will accommodate. This conjecture is the more likely to be true, because sandy spots are always chosen for this purpose, where twenty or thirty burrows are often sunk in close proximity to each other. Such spots would be peculiarly suitable to the Fox, because the sandy soil is not so likely to be hardened by the frost as that of a more compact and watery nature, and would be easily thrown out by the small though powerful feet of the animal.

If one of these little colonies could be laid open, a very curious sight would present itself. The earth would be seen to be pierced with multitudinous tunnels, each complete and independent in itself, and never interfering with burrows belonging to other owners. Each burrow, too, is of a very complex character, and by no means consists of a single tunnel, with a rude nest at the extremity. There are three or four distinct passages, each of which opens into the common chamber, which is of considerable dimensions, and serves as a starting-place whence the inhabitant can seek refuge in either of its passages, according to the direction in which it apprehends danger.

This chamber is not, however, the nursery for the young, a second cavity being used for that purpose. The nursery is not of great dimensions, and communicates by a passage with the chamber already mentioned. The reader will see, therefore, that in some respects the habitation of the Arctic Fox corresponds with that of the mole, both having a kind of fortress from which a number of passages lead in different directions, and the nursery being in both instances separate from the general habitation.

Five or six young ones are mostly bred in these subterranean nurseries; and in the outer chamber, and in several of the passages that lead to it, are placed good stores of food. In one such nest were found many bodies of two species of lemming, and several stoats; and the abundance of bones belonging to hares, fishes, and ducks, showed that the wants of the young Foxes had been amply supplied.

THE habitation of the common Fox of this country is by no means so complicated as that of the Arctic species.

Whenever it can, the Fox avoids the labour of burrowing, and avails itself of the deserted home of a badger, or even a rabbit. In the former case there is very little to be done to the burrow, and in the latter the cunning animal finds its labour greatly diminished; for though the Fox is a much larger animal than the rabbit, and needs a rather larger tunnel, it finds that the task of enlarging a ready-made burrow is very much less than if it had to drive a passage through solid ground. Every one who has worked with carpenters' tools knows that a large gimlet passes easily through wood, if it follows the track of a smaller one, and on the same principle, the Fox passes easily through the earth on the track of the rabbit. The burrow of the latter animal is moreover much larger than is absolutely required for its passage, while the former is quite satisfied if he can pass through the tunnel with tolerable rapidity.

Sometimes, however, the animal is not fortunate enough to find any ready-made habitation, and in such cases sets determinately to work, and scoops out a burrow on its own account. Herein it lies asleep all day, as is the custom with most predaceous animals, and only sallies forth at night. Herein the mother produces and nurtures her young, and sometimes on a summer's evening, the whole family, the father, mother, and cubs, come out to enjoy the fresh air. They never wander far from the mouth of the burrow, and as the young are gamesome little creatures, as playful as puppies, and much prettier, and the mother helps her young ones in their sports as a good mother ought to do, the group presents a very pretty sight. When young the cubs are certainly not prepossessing, and scarcely any one would take the sprawling grey-coated, broad-muzzled creatures, with their little short pointed tails and stumpy ears, for the young of the Fox, with its ruddy fur, its active limbs, its narrow muzzle, its full bushy tail, and its erect, intelligent-looking ears.

THE Weasels have been said to be great burrowers, but I am inclined to think that very few of them are in the habit of tunnelling below the ground.

One of the Weasel tribe is, however, a most powerful and industrious excavator. This is the BADGER (*Meles taxus*), an animal which was formerly considered as our only surviving British representative of the bear tribe, but is now found to belong to the weasels.

The Badger makes a most gloomy, dark, and tortuous burrow, generally excavated in some retired and shadowy spot, such as dense thickets, or the recesses of thickly-wooded forests. As is the case with several burrowing animals, there are several chambers in its domicile, one of which is appropriated as a nursery, and is warmly padded with dry mosses and grass.

The Badger is a creature that cannot live in close proximity to human beings, and has, in consequence, been gradually banished from the greater part of England. Forest after forest falls before the woodman's axe, mile upon mile of barren bog-land is drained and converted into fertile, food-producing soil; and so, to the very great satisfaction of the political economist, and the very great discomfiture of the naturalist, all our large carnivora, whether furred or feathered, are gradually ousted from the soil whereon they formerly exercised unquestioned sway. The Badger has long ago been driven out of the land; the otter is but seldom seen in the rivers where it was once so plentiful; the polecat and martens have retired into the deepest recesses of the few forests which are still left to us, but over which the demon of bricks and mortar already casts an evil eye; and the stoat and weasel only hold their own on account of their diminutive size, and the comparative ease with which they obtain a supply of food. They are among the animals which are gradually eliminated out of existence by the encroachments of man, and it may be that in a few years a stoat or weasel may be as rare in England as a Badger is at the present day.

THE exact classification of animal habitations involves a task not easily accomplished, inasmuch as so many of them partake of characteristics which might entitle them to be placed under various categories. The rabbit, for example, might be considered either as a social or a burrowing animal, and the same may be said of the common wasp, the humble bee, and many other insects.

The PRAIRIE DOG (*Spermophilus Ludovicianus*) may, like the rabbit, be considered equally as a burrower or a social animal, and we will therefore place it in the former of these categories.

This animal is sometimes called the WISH-TON-WISH, but it is usually known by the name of Prairie Dog, though it is a rodent and not a carnivorous animal. The reason of its popular name lies in the short yelping sound which it is fond of uttering, and which bears some resemblance to the bark of a young puppy. Even in captivity it utters this short, impatient yelp, which may generally be extorted from the little animal by placing the hand near the cage.

In spite of the formidable foes by which it is attacked, and which take up their residence in the very centre of its habitations, the Prairie Dog is an exceedingly prolific animal, multiplying rapidly, and extending its excavations to vast distances. Indeed, when once the Prairie Dogs settle themselves in a convenient spot, their increase seems to have no bounds, and the little heaps of earth which stand near the mouth of their burrows extend as far as the eye can reach.

The burrows are of considerable dimensions, and evidently run to no small depth, as one of them has been known to absorb five barrels of water without being filled. It is not impossible, however, that there might have been a communication with some other burrow, or that the soil might have been loose and porous, and suffered the water to soak through its substance. They are dug in a sloping direction, forming an angle of about forty-five degrees with the horizon, and after descending for five or six feet, they take a sudden turn, and rise gradually upwards. Thousands upon thousands of these burrows are dug in close proximity to each other, and honeycomb the ground to such an extent that it is rendered quite unsafe for horses.

The scene presented by one of these 'dog towns' or 'villages,' as the assemblages of burrows are called, is most curious, and well repays the trouble of approaching without alarming the cautious little animals. Fortunately for the traveller, the Prairie Dog is as inquisitive as it is wary, and the indulgence of its curiosity often costs the little creature its life. Perched on the hillocks which have already been mentioned, the Prairie Dog is able to survey a wide extent of horizon, and as soon as it sees an intruder, it gives a sharp yelp of alarm, and dives into its burrow, its little feet knocking together with a ludicrous flourish as it disappears. In every direction a similar scene is enacted. Warned by the well-known cry, all the Prairie Dogs within reach repeat the call, and leap into their burrows. Their curiosity, however, is irrepressible, and scarcely have their feet vanished from sight, than their heads are seen cautiously protruded from the burrow, and their inquisitive brown eyes sparkle as they examine the cause of the disturbance.

The Prairie Dog has not the privilege of possessing a home exclusively devoted to its own use, for the Burrowing Owl, sometimes called the Coquimbo Owl (*Athene cunicularia*), and the terrible rattlesnake, take forcible possession of the burrows, and devour the inmates, thus procuring board and lodging at very easy rates. The rattlesnake at all events does so, the bodies of young Prairie Dogs having been found in its stomach.

On the discovery of owls and rattlesnakes within the burrows of the Prairie Dog, it was generally thought that these incongruous beings associated together in perfect harmony, forming in fact a 'Happy Family' below the surface of the ground. The ruthless scalpel of the naturalist, however, effectually dissipated all such romantic notions, and proved that the snake was by no means a welcome guest, but an intruder on the premises, self-billeted on the inmates, like soldiers on obnoxious householders, procuring lodging without permission, and eating the inhabitants by way of board.

The reason for the presence of the owls is not so evident, though it is not impossible that they may also snap up an occasional Prairie Dog in its earliest infancy, while it is very young, small, and tender. These winged and scaled intruders are not found in all the burrows, though many of the habitations are infested by them.

The general aspect of the Prairie Dog is not unlike that of its near relative, the Alpine Marmot, so familiar in this country through the mediumship of Savoyard boys, who carry the animal about in a box, and exhibit it for halfpence.

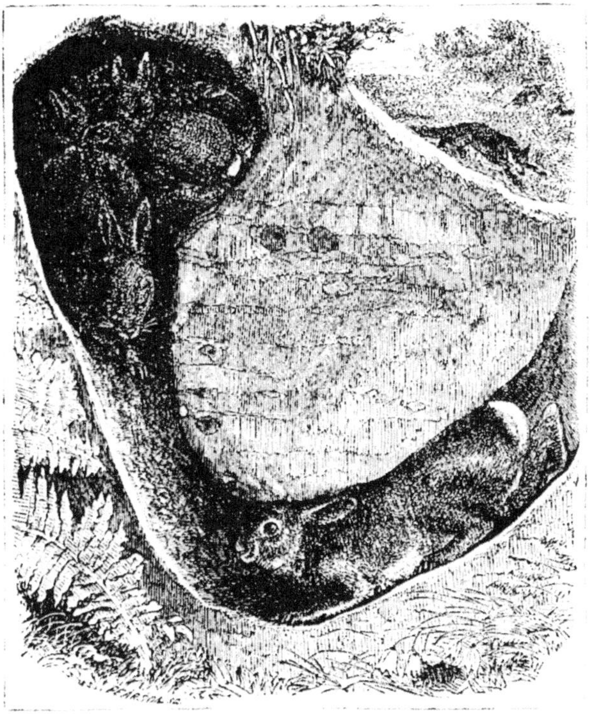

THE RABBIT WARREN.

One of the most familiar of the British burrowing rodents, is the common Rabbit (*Lepus cuniculus*), an animal notable for 'sporting,' as gardeners would say, into a vast number of varieties, some of which are so unlike the original stock, that they seem to be species and not varieties, and indeed might have taken rank as species, did they not invariably display a tendency to recede to the ancestral short brown fur and upright ears of the wild Rabbit.

The animal lives, as we all know, in burrows, and is mostly of a social nature, a considerable number of burrows being gathered together and known by the name of a Warren. Whenever the Rabbits find an undisturbed spot, which combines the advantages of a sandy situation with the vicinity of food, they establish themselves forthwith, and sink their multitudinous tunnels into the ground. The favourite locality for the Rabbit is a loose, sandy, or gravelly soil, covered with patches of furze bushes; for the soil is easily excavated, and is very dry, and the young shoots of the furze yield a food equally grateful and nutritious. Moreover, the tangled roots of the furze afford an excellent protection to the burrows, and the overhanging branches, with their prickly verdure, serve admirably to shelter the entrances.

As is the case with most animals, the Rabbit seeks a quiet and retired spot for her little nursery. She does not produce her young in any of the burrows to which the general Rabbit colony has access, but prepares an isolated tunnel, at the end of which she forms her nest. The bed on which the young recline is beautifully soft and fine, being composed chiefly of the downy fur which grows on the mother's breast, and which she plucks off with her teeth in tufts of considerable size. Any one who keeps tame Rabbits may see the female preparing her cradle with this soft fur, and note how perseveringly she denudes her breast of its covering.

North America is peculiarly rich in burrowing animals belonging to this order—so rich, indeed, that many curious species must be omitted for lack of space.

Among these burrowers, the Chipping Squirrel, or Hackee, or Chipmuck (*Tamias Lysteri*), is peculiarly conspicuous. It is a very pretty little creature, brownish grey in colour, with five stripes of black and two of pale yellow drawn along the back; so that it cannot be mistaken for any other animal. Below, and on the throat, it is a pure snowy white. These are the normal hues of the fur; but it is somewhat variable in point of colour, the grey and yellow being sometimes quite superseded by the black.

The burrow of the Chipping Squirrel is rather complicated in structure, and is always made under the shelter of a wall, an old tree, or a bank. The hole descends almost perpendicularly for nearly a yard, and then makes several devious windings in a slightly ascending direction. Two or three supplementary galleries are driven from the principal burrow, and by means of them the animal is able to escape almost any foe The stoat, however, cannot be deceived by this complicated arrangement of tunnels, but winds its lithe body through all the deviating passages, and kills every Chipping Squirrel which it finds. One of these bloodthirsty weasels has been known to enter the burrow of a Chipping Squirrel, and in a short time to leave it, having in the space of a very few minutes killed six victims, a mother and five young, whose lifeless bodies were found in the nursery when the burrow was opened.

The nest is made of dried leaves of various kinds, and in it the mother and her offspring can rest in security from all ordinary foes. Owing to the complexity of the burrow, no little skill is required to trace its various windings, and much exertion is needed before they can all be laid bare.

OUR next example is the CANADA POUCHED RAT (*Pseudostoma bursarius*), sometimes called the GOPHER, or MULO.

This remarkable animal drives burrows of very great extent, and whenever it gains admission into a garden, it works much damage to the roots of the plants. Every root that crosses the tunnel the Pouched Rat will eat; and not only herbs and flowers, but even fruit trees of many years' growth have been killed by this destructive animal. In such cases, the extremity of the burrow is always to be found among the roots of some tree, which act at the same time as a defence and a larder; for the Rat hides itself under their protection, and eats away their tender shoots.

Like the mole, the Gopher throws up little hillocks at irregular intervals, sometimes twenty or thirty feet apart, and sometimes crowded closely together. The nest of the Gopher is made in a burrow constructed expressly for the purpose, and is placed in a small globular chamber about eight inches in diameter. The bed on which the mother and her young repose is made of dried herbage and fur plucked from the body. This chamber is the point from which a great number of passages radiate, and from these other tunnels are driven. These radiating burrows evidently serve two purposes, enabling the animal to escape in any direction when alarmed, and serving to conduct it to its feeding grounds.

CANADA POUCHED RAT.
(*Plan of Burrow.*)

The Canada Pouched Rat is nearly a foot in length, and is notable for the great development of its incisor teeth, which project beyond the lip : and for the dimensions of its cheek-pouches, which measure about three inches in length, and extend as far as the shoulders. It was formerly thought that the animal employed these pouches for the conveyance of earth out of its burrow, but it is now known that it does not make any such use of these natural pockets.

OWING to the peculiar nature of the substance in which the WHITE BEAR (*Thalarctos maritimus*) makes its curious burrows, I have placed it after, instead of before, the earth-burrowing rodents.

POLAR BEAR.

Towards the month of December the WHITE BEAR retreats to the side of a rock, where by dint of scraping, and allowing the snow to fall upon her, she forms a cell in which to reside during the period of her *accouchement.* Within this strange nursery she produces her young, and remains with them beneath the snow until the month of March, when she emerges into the outer air, bringing with her the baby bears, who are then about as large as ordinary rabbits. As the time passes on, the breath of the family, together with the warmth exhaled from their bodies, serves to enlarge the cell, so that in proportion with their increasing dimensions, the accommodation is increased to suit them. Although covered so deeply, the hidden Bear may be discovered by means of the little hole which is made by the warm breath, and is rendered more distinguishable by the hoar-frost which collects around it.

This curious abode is not sought by every Polar Bear. None of the males trouble themselves to spend so much time in a state of seclusion; and as the only use of the retreat is to shelter the young, the unmarried females roam freely about during the winter months. The habit of partial hibernation is common to most, if not to all true Bears, and we find that the White Bear of the Polar regions, the Brown Bear of Europe, and the Black Bear of Northern America, agree in this curious habit. Before retiring into winter quarters, the Bear eats enormously, and, driven by an unfailing instinct, resorts to the most nutritious diet, so that it becomes prodigiously fat. In this condition it is in the best state for killing, as the fur partakes of the general fulness of the body, and becomes thick and sleek, as is needful when we consider the task which it has to perform.

During the three months of her seclusion, the Polar Bear takes no food, but exists upon the store of fat which has been accumulated before retiring to her winter home. A similar phenomenon may be observed in many of the hibernating animals, but in the Bear it is more remarkable from the fact that she has not only to support her own existence, but to impart nourishment to her cubs. It is true, that in order to enable them to find sufficient food, they are of wonderfully small dimensions when compared with the parent; but the fact remains, that the animal is able to lay up within itself so large a store of nutriment that it can maintain its own life and suckle its young for a space of three months without taking a morsel of food.

FROM a work of this character, so remarkable an animal as the PICHICIAGO ought not to be omitted. Its scientific name is *Chlamyphorus truncatus*, and is very properly chosen, as will presently be seen.

PICHICIAGO.

The Pichiciago is not larger than an ordinary mole, and in its general habits somewhat resembles that animal. The shape of its body sufficiently indicates its burrowing propensities, and the view of the skeleton confirms the aspect of its outward form. The bones of the fore legs are short, thick, and arched in that manner which is so indicative of great muscular power, and even those of the hind legs are remarkably strong in proportion to the size of the animal. The fore paws are enormously large, palm-shaped, and furnished with five strong, curved, and compressed claws, so as to form admirable digging instruments. The snout is rather long and pointed, and, as in the mole, the eyes are very small, and hidden under the soft dense fur.

It is a native of Chili, and seems to be of rare occurrence, though it may probably be more plentiful than is generally imagined, its subterranean habits and timid nature seldom permitting it to be seen. Like the mole, it lives beneath the earth, scooping out long galleries in the soil, and probably feeding upon insects, worms, and grubs like the rest of the edentate animals.

The chief point of interest which strikes an observer when looking at a Pichiciago, is the cuirass with which its body is defended. It is made and arranged in a very peculiar manner. The cuirass looks as if a number of squared plates of horn had been sewn upon short lengths of tape, and then the tape bands laid side by side and fastened to each other. It is not fixed to the animal throughout its whole extent, as might be supposed, but is only attached along the spine, and on the top of the head. It does not merely protect the back, but when it reaches the insertion of the tail, turns suddenly downwards as if on hinges, and forms a kind of flap over the hind-quarters, which are short and square, as if abruptly cut off by a perpendicular blow with a sharp instrument. This arrangement affords a perfect protection to the hind-quarters while the animal is burrowing, and effectually repels any attack that might be made from the rear, reminding the observer of the shell with which the testacella is furnished.

This coat of mail is as flexible as the chain or scale armour of the olden times, and accommodates itself to every movement of the animal. The rest of the body is covered with a coat of soft, yellowish fur, nearly as fine as that of the mole, and much longer, but not so dense. The scientific name of the Pichiciago relates to the mail-clad body and the peculiar form of the hind-quarters, the generic title signifying 'mantle-bearer,' and the specific name, 'abruptly shortened.'

THE different species of Manis deserve a passing notice. They are all burrowers, and are furnished with armour even better calculated for defence than that of the armadillo, inasmuch as it assumes somewhat of an offensive as well as a defensive character. All these animals are covered with large, sharp-edged scales, of a stout horny consistence, which overlap each other, like the tiles of a house. They are of wonderful hardness, and form a buckler which is impenetrable to any weapon possessed by the carnivorous animals of the regions wherein it resides. A specimen of the BAJJERKEIT, or SHORT-TAILED MANIS of India (*Manis pentadactyla*), now before me, affords a good example of this weapon-resisting power. Edward Arnold, Esq., to whom I am indebted for this specimen, possessed it in a living state for a considerable time, and, when he was about to leave India, determined to kill the animal and take the skin with him. Accordingly, he fired three barrels of a Colt's revolver pistol at the Manis, but without the slightest effect, and was at last obliged to introduce the point of a dagger under the scales, and drive the weapon into the heart. On examining the interior of the skin, the wound caused by the double-edged dagger is plainly perceptible, but I cannot find the slightest trace of the bullets. One of the balls, indeed, recoiled upon the intending destroyer.

When the Manis is alarmed, it rolls itself up, wraps its tail over the body, and lies in conscious security, the horny scales acting as a buckler, and their sharp edges deterring enemies from the attack as much as the quills of the porcupine or the spines of the hedgehog.

THE curious AARD VARK of Southern Africa (*Orycteropus Capensis*) is another of the earth-burrowers, residing, for the most part, in great holes which it scoops in the ground.

The name Aard Vark is Dutch, signifying Earth-hog, and is given to the animal on account of its extraordinary powers of excavation and the swine-like contour of its head. The claws with which this animal works are enormous, as, indeed, is needful for the task which they are intended to perform. They are by no means intended merely to excavate burrows in soft or sandy soil, though they are frequently employed for that purpose, but they are designed for labours far more arduous. By means of these implements, the Aard Vark tears to pieces the

AARD VARK.

enormous ant-hills which stud the plains of Southern Africa—edifices so strongly made as to resemble stone rather than mud, and capable of bearing the weight of many men on their summits. These marvellous dwellings (of which we shall see something in a future page) are absolutely swarming with inmates; and it is for the purpose of feeding upon the tiny builders that the Aard Vark plies its destructive labours.

Towards evening the Aard Vark issues from the burrow wherein it has lain asleep during the day, proceeds to the plains, and searches for an ant-hill in full operation. With its powerful claws it tears a hole in the side of the hill, breaking up the stony walls with perfect ease, and scattering dismay among the inmates. As the ants run hither and thither, in consternation, their dwelling falling like a city shaken by an earthquake, the author of all this misery flings its slimy tongue among them, and sweeps them into its mouth by hundreds Perhaps the ants have no conception of their great enemy as a fellow-creature, but look upon the Aard Vark as we look upon an earthquake, the plague, or any other disturbance of the usual routine of nature. Be this as it may, the Aard Vark tears to pieces many a goodly edifice, and depopulates many a swarming colony, leaving a mere shell of irregular stony wall in the place of the complicated and marvellous structure which had sheltered so vast a population.

THERE are two large islands, one large enough to take rank as a continent, which are pre-eminent for the strange character of the creatures which inhabit them. Whenever an animal of more than usual oddity is brought to England, we may safely conjecture that it was taken either in Madagascar or Australia. The creatures which we are now about to examine are natives of the latter country.

Perhaps there never was a more extraordinary and unique being than the well-known animal which is so familiar to us under many titles. Some call it the DUCKBILL, on account of its mandibles, which are ludicrously like those of the bird from which it derives its name. Others call it the WATER MOLE, on account of its aquatic habits and mole-like fur.

Some scientific naturalists have called it the *Ornithorhynchus paradoxus*; others have given it the name of *Platypus anatinus* —the former title being to my mind by far the more appropriate and expressive of the two. The vatives [illegible] trali have several names for this remarkable animal; so [illegible] alling [illegible] Mallangong, others Tambreet, and others Tohunb[illegible]k—the second of these titles being most generally in use.

On looking at a living Duckbill, few would set it down as an excavator of the soil; yet it is a burrower, and makes tunnels of great length and some complexity. The soft broad membrane that extends beyond the claws while the animal is walking or swimming, and in the latter case forms a paddle by which the creature can propel itself swiftly through the water, falls back when the foot is employed for digging, and aids the animal in flinging back the soil which its claws have scraped away. The round body is admirably adapted for traversing the burrows, though the stuffed specimens which generally are seen in museums give but little idea of such capability. As a general rule, these stuffed specimens are much too long, too stiff, too straight, too flat, and too shrivelled. During life, the body is round, and the skin hangs in loose folds around it, having a very curious aspect when the creature is walking upon the land. The Duckbill is, in fact, so very odd a being, that dogs who see it for the first time, as it scrambles along with its peculiar waddling gait, will sit and prick up their ears, and bark at the strange animal, but will not dare to meddle with it; while cats fairly turn tail, and scamper away from so uncanny a beast. The hair with which the body is so densely covered is admirably suited to an animal which passes its time in the water or underground. Next the skin there is a thick close coating of woolly fur, through which penetrates a second coat of long hairs, which are very slender at their bases, and can therefore turn in any direction, like those of the mole. The eyes are fuller and rounder than might be expected in an animal that passes so much of its time underground; but they are defended from the earth by a remarkable leathery flap, which surrounds the base of the mandibles, and looks very like the leathern guard of a foil. This curious appendage has probably another use, and is intended to prevent the bill from being thrust too deeply into the mud when the animal is engaged in searching for food.

The wonderful duck-like mandibles into which the head is prolonged are sadly misrepresented in the stuffed specimens which we generally see, and are black, flat, stiff, and shrivelled, as if cut from shoeleather. No one would conceive, after inspecting a dried specimen, how round, full and pouting were once those black and wrinkled mandibles, and how delicately they had been coloured while the animal retained life. Their natural hue is rather curious, the outer surface of the upper mandible being very dark grey, spotted profusely with black, and its lower surface pale flesh-colour. In the lower mandible the inner surface is flesh-coloured, and the outer surface pinky white, sometimes nearly pure white.

Having now glanced at the general form of the Duckbill as it is in life, and not as it is in museums, we will pass to the habitation which it constructs.

Being a peculiarly aquatic animal, the Duckbill always makes its home in the bank of some stream, almost invariably at those wider and stiller parts of the river, which are popularly called ponds. There are always two entrances to the burrow, one below the surface of the water, and the other above, so that the animal may be able to regain its home either by diving, or by slipping into the entrance which is above the surface. This latter entrance is always hidden most carefully under overshadowing weeds and drooping plants, and is so carefully concealed that the unaccustomed eyes of an European can very seldom find it.

When the grasses, &c. are put aside, there is seen a hole of moderate size, on the sides of which are imprinted the footmarks of the animal. By the dampness and sharpness of these impressions, the natives can form a tolerably accurate opinion whether the creature is likely to be at home or not, as in the former case, the footmarks which point upwards are fresher and wetter than those which point downwards. While digging out the Duckbill, they occasionally pull out a handful of the clay, inspect the marks, and then fall to work afresh. From this hole the burrow passes upwards, winding a sinuous course, and often running to a considerable length. From twenty to thirty feet is the usual average, but burrows have been opened where the length was full fifty feet, and where the course was most annoyingly variable, bending and twisting about so as to tire the excavators, and make them quite disgusted with their work. The natives never dig out the entire burrow, but push sticks along it, and sink shafts upon the sticks; just, in fact, as a boy digs out a humble bee's nest, by inserting twigs into the hole, and digging down upon them.

This serpentine form of burrow is in all probability attributable in a great degree to the peculiar instincts of the animal. As, however, the course of the tunnel is extremely variable, and no two burrows have precisely the same curves and windings, it is likely that various obstacles, such as roots and stones, may turn the animal out of its course while engaged in digging its subterranean home, and therefore that the shape of the burrow may in some degree depend upon the character of the ground.

At the upper extremity of the burrow is placed the nest, an excavation of a somewhat oval form, much broader than the width of the burrow, and well supplied with dry weeds and grasses, upon which the young may rest. They appear to remain in these burrows until they have attained half their full growth, for Dr. Bennett captured a pair of young Ducklings, ten inches in length, which seemed not to have left the burrow. Sometimes there are four young in one nest, and sometimes there is only one, but the usual number is two.

THERE is another strange Australian animal, also remarkable for its power of burrowing. This is the creature which is known as the PORCUPINE ANT-EATER (*Echidna hystrix*), and is called by the very erroneous names of Porcupine, or Hedgehog. The natives have several names for it, some calling it Nicobejan, others Jannocumbine, and others Cojera.

The Echidna is a wonderful burrower, and, in spite of its small size, can make its way through very hard ground. It can pull up stones of great size if it can only contrive to insert its paws and find a convenient crevice for them, and is so quick at this task that to confine the animal is by no means an easy matter, even a paved yard affording but a poor safeguard against its escape. In the open country it digs with such extreme rapidity that it can hardly be captured, gathering its back into an arched form, collecting the legs under the body, scratching away with the feet, and sinking like a stone in a cup of treacle.

If attacked when on ground into which it cannot burrow rapidly, the Porcupine Ant-eater immediately turns itself into a ball, hedgehog-wise, and sets its foes at defiance. The large perforated spur with which the hind feet of the male are armed. and through which is poured a liquid secreted by a gland of considerable size, is a very formidable-looking weapon, but to all appearances is really harmless. Dr. Bennett often handled the animal, but never saw it attempt to use the spur, and found that the duckbill, which is armed in a similar manner, was equally innocuous.

At the present date, January, 1864, the living animal may be seen in the collection at the Zoological Gardens.

CHAPTER II.

BURROWING BIRDS.

The SAND MARTIN—Mode of burrowing and shape of the tunnel—Enemies of the Sand Martin—Midges and Martins—The KINGFISHER and its habits—Its burrow and peculiar nest—Number of the eggs—The PUFFIN a feathered usurper—The Feroe Islands and the Puffins—Pro aris et focis—The JACKDAW, STOCKDOVE, and SHELDRAKE—Nest of the Sheldrake—The STORMY PETREL—Its mode of nesting and shallow tunnels—mode of feeding its young—Evil odour of its burrow—The WOODPECKER—Its uses and misunderstood character—Method of burrowing—The Fungus and the Woodpecker.

WE now take leave of the furred burrowers, and proceed to those which wear feathers instead of hair.

One of the best examples of Bird Burrowers is the well-known SAND MARTIN (*Cotile riparia*), so plentiful in this country. The powers of this pretty little bird seem to be quite inadequate to the arduous labours which it performs so easily, and few would suppose, after contemplating its tiny bill, that it was capable of boring tunnels into tolerably hard sandstone. Such, however, is the case, for the Sand Martin is familiarly known to drive its tunnels into sandstone that is hard enough to destroy all the edge of a knife.

The bird does not prefer a laborious to an easy task, and if it can find a spot where the soil is quite loose, and yet where the sides of the burrow will not collapse, it will always take advantage of such a locality. I have frequently seen such instances of judgment, where the birds had selected the sandy intervals between strata of stone, and so saved themselves from any trouble except scraping and throwing out the loose sand.

When, however, the Sand Martin is unable to find such a situation, it sets to work in a very systematic fashion, trying several successive spots with its beak, until it discovers a suitable locality. It then works in a circular direction, using its legs as a pivot, and by dint of turning round and round, and

SAND MARTINS.

pecking away as it proceeds, soon chips out a tolerably circular hole. After the bird has lived for some time in the tunnel, the shape of the entrance is much damaged by incessant passing to and fro of the inmates, but while the burrow is still new and untenanted, its form is almost cylindrical. In all cases the tunnel slopes gently upwards, so as to prevent the lodgement of rain, and its depth is exceedingly variable. About two feet and a half is a fair average length. Generally, the direction of the burrow is quite straight, but sometimes it takes a curve, where an obstacle, such as a stone or a root has interrupted the progress of the bird. Should the stone be a large one, the Sand Martin usually abandons the burrow, and resumes its labours elsewhere, and in a piece of hard sandstone rock many of these incomplete excavations may be seen.

At the furthest extremity of the burrow, which is always rather larger than the shaft, is placed the nest—a very simple structure, being a little more than a mass of dry herbage and soft feathers, pressed together by the weight of the bird's body. Upon this primitive nest are laid the eggs, which are very small, and of a delicate pinky whiteness.

Few foes can work harm to the Sand Martin, during the task of incubation. Rats would find the soft sandy soil crumble away from their grasp ; and even the lithe weasel would experience some difficulty in gaining admission to the nest. After the young Sand Martins are hatched, many foes are on the watch for them. The magpie and crow wait about the entrance of the holes, in order to snap up the inexperienced birds while making their first essay at flight; and the kestrel and sparrow-hawk come sweeping suddenly among them, and carry off some helpless victim in their talons.

Man is perhaps the worst foe of the Sand Martin, for there is a mixture of adventure and danger in taking the eggs, which is irresistible to the British schoolboy.

Fortunately for the Sand Martins, many of their nests are placed in situations which no boy can reach, and there are happily some instances where the services which they render to mankind are properly appreciated. Mr. C. Simeon, in his 'Stray Notes on Fishing and Natural History,' gives an interesting account of some Sand Martins which are thus gratefully protected :—

'Whilst waiting for the train one afternoon at Weybridge, I

amused myself with watching the Sand Martins, who have there a large establishment on either side of the cutting, and got into conversation with one of the porters about them. On my saying, I supposed that the boys robbed a good many of the nests, he answered, "Oh, sir, they would if they were allowed, but the birds are such good friends to us, that we won't let anybody meddle with them." I fancied at first that he spoke of them as friends in the way of company only, but he explained his meaning to be, that the flies about the station would be quite intolerable if they were not cleared off by the martins, which are always hawking up and down in front of it; adding, that even during the few hot days which occurred in the spring before their arrival, the flies were becoming very troublesome. "Now," he said, "we may now and then see one, but that is all."

'It was a bright sunny day in July, and the scene was a very lively and interesting one. The mouths of the holes on both sides of the cutting were crowded with young martins—as many perhaps as four or five in each—sunning their barred white breasts, and waiting to be fed: the telegraph wires formed perches, of which advantage was taken by scores of others more advanced in growth, and of old ones reposing after their exertions; while the air was filled with others employed in catering for their families. All of a sudden the young ones retreated into their holes; the wires were deserted, and only a few remained, describing distant circles. I thought that a hawk must have made his appearance, but it turned out that the alarm had been caused by two men walking over the heath above, and approaching the holes. The young ones in the holes had, no doubt, felt the jar caused by their tread, and those on the wing, who saw them, had probably given warning, by note, to the others perched on the wires, who could not have seen, nor, I should think, heard their approach.'

ALTHOUGH THE KINGFISHER (*Alcedo ispida*), does not excavate the whole of the burrow in which it resides, it does, at all events, alter and arrange a ready-made burrow to suit its own necessities.

Generally, the nest is placed in the deserted burrow of a water-vole, but in this instance it had been made in the empty tunnel of a water-shrew, so that the hole was of comparatively small dimensions, and would not admit my hand and arm without some artificial enlargement. In all cases, the bird takes care to increase the size of the burrow at the spot where the nest is made, and to choose a burrow that slopes upwards, so that however high the water may rise, the nest will be perfectly dry.

That the eggs are laid upon dry fish-bones is a fact that has long been known, but for an accurate account of the nest we are indebted to Mr. Gould, the eminent ornithologist.

Until he succeeded in removing the nest entire, no one had been able to perform such a feat, and so well known to all bird-nesters is the difficulty of the task, that a legend was, and perhaps is still, current in various parts of England, that the authorities of the British Museum had offered a reward of 100*l.* to anyone who would deposit in their collection a perfect nest of the Kingfisher. This feat has been admirably accomplished by Mr. Gould.

The nest is composed wholly of fish-bones, minnows furnishing the greater portion. These bones are ejected by the bird when the flesh is digested, just as an owl ejects the pellets on which her eggs are laid. The walls of the nest are about half an inch in thickness, and its form is very flat. The circular shape and slight hollow show that the bird really forms the mass of bones into a nest, and does not merely lay her eggs at random upon the ejecta. The whole of these bones were deposited and arranged in the short space of three weeks.

It may possibly be owing to these bones and the partial decomposition which must take place during the time occupied in drying, that the burrow possesses so exceedingly evil an odour. This unpleasant effluvium, which may indeed be called by the stronger name of stench, is wonderfully enduring, and clings to the bird as well as to its dwelling. The feathers of the Kingfisher are most lovely to the eye, but the proximity of the bird is by no means agreeable to the nostrils, the 'ancient and fish-like smell' being extremely penetrating. I have now before me a stuffed and perfectly dry skin of a Kingfisher, which has been washed and soaked in water for many hours, and yet retains the peculiar odour, which is so strong that after I had prepared it, many and copious ablutions were required to divest my hands of the horrible emanation.

To those who collect eggs, and care for numbers, the discovery of a Kingfisher's nest is a singular boon. Not only does the bird lay a great multitude of eggs, the aggregate mass of which exceeds her own dimensions, but she is a fearless and indefatigable layer, and if the eggs are removed with proper care, she will produce an enormous number in the course of a season.

THE comical little PUFFIN (*Fratercula arctica*) may be reckoned among the true burrowers, possessing both the will and the power of excavation, but exercising neither unless pressed by necessity.

As is the custom with most diving birds, the Puffin lays only one egg, and always deposits it in some deep burrow. If possible, the bird takes advantage of a tunnel already excavated, such as that of the rabbit, and 'squats' upon another's territory, just as the Coquimbo owl takes possession of the excavations made by the prairie dog. The rabbit does not allow its dominion to be usurped without remonstrance, and accordingly the bird and the beast engage in fierce conflict before the matter is settled. Almost invariably the Puffin wins the day, its powerful beak and determined courage being more than a match for the superior size of its antagonist.

When it is unable to obtain a ready-made habitation, it sets to work on its own account, and excavates tunnels of considerable dimensions.

The Feroe Islands are notable haunts of the Puffin, because the soil, which is in many places soft and easily worked, is favourable for its excavations. The male is the principal excavator, though he is assisted by the female; and so intent is the bird upon its work, that it may be captured by hand by thrusting the arm into the burrow. The average length of the tunnel is about three feet, and it is seldom straight, taking a more or less curved form, and being furnished with a second entrance. No nest of any kind is used, but the egg is laid on the earth, at the end of the burrow, so that, although it is at first beautifully white, it becomes in a short time stained so deeply that it can seldom be restored to its primitive purity.

So deeply do the burrows run, that when a passenger is walking near the edge of the precipice upon which the Puffins

PUFFIN.

breed, he can hear the old birds grunting below his feet, angry because they are disturbed by the footsteps above them.

The young Puffin has many foes, who endeavour to seize it before its bill has attained its full proportions and its muscles have gained their full powers. The parent birds, however, bravely defend their young, and have been known, as a last resource, to grasp the invader in the beak, and hurl themselves and the foe into the sea. Once among the waves, the Puffin is in its natural element, for it is an admirable swimmer and practised diver, being able to catch the swift-finned fishes and bear them home to its nest. The foe, therefore, must either remain on dry land or lose the victory, if not its life, for there are few enemies for which the Puffin is not more than a match when in the water.

THERE are many other birds which pass a semi-burrowing life, making their nests in hollows already excavated, and either using them without adaptation or altering them very slightly for the purpose of depositing their eggs and rearing their young. The JACKDAW, for example (*Corvus monedula*), is frequently one of the semi-burrowers, making its nest within deserted rabbit burrows, when it can find no more congenial locality. The STOCKDOVE (*Columba œnas*) is frequently found in similar situations, placing its rude platform of sticks within the burrow; and the common SHELDRAKE (*Tadorna Vulpanser*) possesses the same habit.

The nest of the last-mentioned bird is always placed close to the water, so that the young may be fed with marine crustacea. The female is accustomed to cover the eggs with down plucked from her own breast. Rabbit warrens upon sea-edged cliffs, are favourite resorts of the Sheldrake. In default, however, of rabbit burrows, the Sheldrake is well content with any moderately deep holes in the shore, and therein lays her enormous deposit of eggs, which are from ten to fifteen in number, and of a white colour. Burrows thus tenanted may be found in many situations, especially on the banks of estuaries, localities which are always sheltered, and almost always produce an abundant supply of food for the bird and its young brood.

WE often find burrowers where we least expect them.

Who would think, on inspecting a specimen of the well-known STORMY PETREL (*Thalassidroma pelagica*), that it was able to dig into the ground, and form the burrow in which it makes its nest? Such, however, is the case, and the pretty little traverser of the ocean shows itself to be as accomplished in excavating the ground as it is in flitting over the waves with its curious mixture of flight and running. If the Stormy Petrel can find a burrow already dug, it will make use of it, and accordingly is fond of haunting rocky coasts, and of depositing its eggs in some suitable cleft. It also will settle in a deserted rabbit-burrow, if it can find one sufficiently near the sea, and is found breeding in many places which would equally suit the puffin.

Failing, however, all natural or ready-made cavities, the Stormy Petrel is obliged to excavate a tunnel for itself, and even on sandy ground is able to make its own domicile. Off Cape Sable, in Nova Scotia, there are many low-lying islands, the upper parts of which are of a sandy nature, and the lower composed chiefly of mud. Not a hope is there in such localities of already existing cavities, and yet to those islands the Petrels resort by thousands, for the purpose of breeding. The birds set resolutely to work, and delve little burrows into the sandy soil, seldom digging deeper than a foot, and, in fact, only making the cavity sufficiently large to conceal themselves and their treasure.

Each bird lays a single egg, which is white, and of small dimensions. The young are funny-looking objects, and resemble puffs of white down rather than nestlings. The parent attends to its young with great assiduity, feeding it with the oleaginous fluid which is secreted in such quantities by the digestive organs of this bird. So large indeed is the amount of oil, that in some parts of the world the natives make the Stormy Petrel into a lamp, by the simple process of drawing a wick through its body. The oil soon rises into the wick, and burns as freely as in any of the really rude and primitive, though ornamental lamps of the ancients.

The Petrel only feeds its young by night, remaining on the wing during the day, and flying to vast distances from the land. Owing to this habit, and its custom of taking to the sea during the fiercest storms, it has long been an object of dread to sailors, whose illogical minds are unable to discriminate between cause and effect, and fancy that the Petrel, or Mother Carey's Chicken, as they call the bird, is the being which, by the exercise of some magic art, calls the storm into existence. They even fancy that the Petrel never goes ashore nor rests; and will tell you that it does not lay its egg in the ground, but holds it under one wing, and hatches it while engaged in flight. To the vulgar mind, everything incomprehensible is fraught with terrors, and so the harmless, and even useful Petrel, is hated with strange virulence.

Throughout the breeding season, the Petrel is indefatigable in search of food, and will follow ships for considerable distances, in hopes of obtaining some of the offal that is thrown overboard by the cook. Even if a cupful of oil be emptied into the water, the Petrel will scoop it up in its bill, and take it home to its young. During the night it mostly remains with its offspring, feeding it, and making a curious grunting noise, something like the croaking of frogs. This noise is continued throughout the night, and those who have visited the great nesting places of the Petrel, unite in mentioning it as a loud and peculiar sound. The ordinary cry is low and short, something like the quacking of a young duck. By day, however, the birds are silent, and only those who keep nightly watch on the ship's deck, can have an opportunity of hearing their chattering cry.

The burrow in which the young Petrel is hatched is extremely odoriferous, the oily food on which the bird lives having itself a very rancid and unsavoury scent; and in consequence of feeding upon this substance, both the habitation and the inmates are extremely offensive to the nostrils. The young bird is at first very helpless, and remains in its excavated home until it is several weeks of age. One of these birds was seen on the Thames in the month of December, 1823, where it attracted some attention, its peculiar mode of pattering over the water causing it to be taken for a wounded land bird, and inducing many persons to go in vain pursuit of the supposed cripple.

THE birds that have hitherto been mentioned are either burrowers into the earth, or adopters of burrows which have been made and deserted by fossorial mammalia. Those which now come before us are burrowers into wood, and either form their tunnels with their own beaks, or adapt to their purposes the excavations made by other creatures, and the hollows formed by natural decay.

The first in order of these birds are necessarily the WOODPECKERS, examples of which are found in most parts of the world. They are easily distinguished from any other birds by the peculiar construction of the beak, the feet, and the tail; the beak enabling them to chip away the bark and wood, the feet giving them the power of clinging to the tree-trunk, and the tail helping to support them in the attitude which gives to their strokes the greatest force. Their beaks are long, powerful, straight and pointed; their feet are formed for grasping, and are set far back upon the body; and their tails are short and stiff, and act as props when pressed against the rough bark.

As is well known, this bird makes its nest in a tunnel which it hollows in the tree, and to a superficial observer might easily be reckoned among the enemies of the forest. If it were to burrow into sound timber, as is often supposed to be the case, it would certainly rank among the deadliest foes of our trees; for in the spots where it still resides, its burrows may be seen in plenty, perforating the trunks and branches of the finest and most picturesque trees. But, in point of fact, none of the British Woodpeckers are able to cut so deep a tunnel into sound and growing wood, and are perforce obliged to choose timber which is already dead, and which has begun to decay.

Sometimes the bird selects a spot where a branch has been blown down, leaving a hollow in which the rain has lodged and eaten its way deeply into the stem. In such places the wood is so soft that it can be broken away with the fingers, or scraped out with a stick; and in many a noble tree, which seems to the eye to be perfectly sound, the very heart-wood is being slowly dissolved by the action of water, which has gained access through some unsuspected hole. Water, when thus admitted to the interior of a tree, fills its centre with decay; and if a perforation be made through the trunk, so as to let out the contained fluid, gallon after gallon of dark brown water will gush forth, mixed with fragments of decayed wood, and betray, by its volume and consistency, the extent of the damage which it has occasioned.

Oftentimes a large fungus will start from a tree, and in some mysterious manner will sap the life-power of the spot on which it grows. When the fungus falls in the autumn, it leaves scarcely a trace of its presence, the tree being apparently as healthy as before the advent of the parasite. But the whole character of the wood has been changed by the strange power of the fungus, being soft and cork-like to the touch. Although the eye of man cannot readily perceive the injury, the instinct of the Woodpecker soon leads the bird to the spot, and it is in this dead, soft, and spongy wood that the burrow is made. Mr. Waterton, who, I believe, was the first to point out this fact, has shown me many examples of the fungus and its ravages among his trees, several fine ash-trees and sycamores having been reduced to mere stumps by the silent operation of the vegetable parasite.

The pickaxe-like beak of the Woodpecker finds no difficulty in making its way through the decayed wood, and thus the bird is enabled to excavate its burrow without very much trouble. The nest itself can scarcely be called by that name, being nothing more than a collection of the smaller chips which have fallen to the extremity of the tunnel while the bird was engaged in the task of excavating. The burrow of the Woodpecker is as unpleasantly odorous as that of the kingfisher. The eggs are pure white.

CHAPTER III.

BURROWING REPTILES.

The REPTILES and their hibernation—The LAND TORTOISE and its winter dwelling—The CROCODILES—Snakes—The YELLOW SNAKE of Jamaica—Its general habits—Its burrowing powers discovered—Presumed method of removing the earth.

THE REPTILES are, as a body, not remarkable for the burrows which they make.

Many of them bore their way into the ground, pass a few months in a state of torpidity, and then push their way out again. But the hole which they make in the earth is scarcely to be called a home, inasmuch as the inhabitant merely enters it as a convenient place wherein it may become torpid, and abandons it as soon as the ordinary functions of the system are restored by the warmth of the succeeding year.

The common Land Tortoise, for example (*Testudo Græca*), is in the habit of slowly digging a burrow with almost painful deliberation, and then concealing itself below the surface of the earth during the cold months of winter. Many Tortoises which have lived in this country have been noticed to perform this act, and I have lately seen a very good example of a burrow which had been sunk amid some strawberry plants, and from which the inmate had just emerged.

Many other reptiles follow a similar course of action. The crocodiles, for example, sink themselves deeply in the mud, and have more than once caused much alarm by awakening out of their hibernation, and protruding their unwelcome snouts from the mud close to the feet of the astonished spectator.

Snakes are accustomed, in like manner, to conceal themselves during the period of their hibernation, resorting to hollow trees, holes in the ground, and similar localities. Labourers while engaged in digging, especially in breaking down banks, frequently unearth a goodly assemblage of snakes, all coiled up in an unsuspected cavity, which they must have entered through the deserted burrow of a mouse or some other little animal. But that a snake should be able to form its own burrow is a feature so remarkable in herpetology, that a single accredited example must not be passed without notice.

In his very interesting work on the natural history of Jamaica, Mr. Gosse gives a curious account of a burrow made by the YELLOW SNAKE (*Chilabothrus inornatus*). This snake is very plentiful in Jamaica, and is perfectly harmless to man, being destitute of poison-fangs, and not reaching a size which would render it formidable to human beings. Its average length, when full-grown, is eight feet. So far, indeed, from being obnoxious to man, it may rank among his best friends, as being a determined foe to rats, feeding largely upon them, and even entering houses in search of its prey. Like the weasel, indeed, of our own country, which feeds mostly on mice and other destructive animals, but occasionally makes a raid upon the fowl-house, the Yellow Snake enters the farmyard, and, instead of eating rats as it ought to do, proceeds to the hen-roosts, and robs them. No less than seven eggs have been found inside a single Yellow Snake, and not a single egg was broken.

There is now (1863) a good specimen in the Reptile-room of the Zoological Gardens of London.

One of these snakes was seen to crawl out of a hole in the side of a yam-hill—*i.e.* a bank of mould prepared for the purpose of growing yams—and when the earth was carefully removed, a large chamber was discovered in the middle of the hill, nicely lined with strips of half-dried plantain leaves, technically called 'trash,' and containing six eggs, all fastened together. Just outside the hole was a heap of loose mould, which had evidently been thrown out when the excavation was made.

The Yellow Snake generally makes its home in the deep spaces between the spurs of the fig or the buttresses of the cotton-tree, and always lines it with 'trash;' but that the creature should be able to excavate a burrow, and throw out the earth, seems almost incredible. How did the snake remove the earth? As the reptile was not seen in the act of excavating, this question could not be precisely answered. Mr. Hill, however, to whom this subject was referred, gave as his opinion that the snake loosened the earth with its snout, and then worked the loose soil out of the hollow by successive contractions of the segments of the abdomen, which would thus 'deliver' the soil after the manner of the Archimedean screw.

The eggs which were found in the chamber were removed, and from one of them, which was opened, was taken a young snake, about seven inches in length.

CHAPTER IV.

BURROWING INVERTEBRATES.

CRUSTACEA.

The LAND CRABS and their habits—The VIOLET LAND CRAB—Its burrows, its combativeness, and its pedestrian powers—The FIGHTING CRAB, why so called—The RACER CRAB of Ceylon—Its burrows and mode of carrying off the soil—The ROBBER CRAB—Its form and general habits—Food of the Robber Crab—A soft bed, and well-stocked larder—The CHELURA, and its ravages among timber—The GRIBBLE and its kin.

THE reader will doubtless perceive that among such a multitude of mammals and birds, each of which has some habitation, it is impossible to give more than a selection of some of the more remarkable examples. Although, therefore, there are many other burrowing and semi-burrowing vertebrates, we must leave the furred, feathered, and scaled tribes, and pass to those which occupy a lower place in the animal kingdom.

Among the Crustacea, there are very many species which form burrows, and which conceal themselves under the sand or mud. As, however, these creatures cannot be said to form their habitations, and the burrows are mostly obliterated by the return of the water, they can scarcely be reckoned among those which make 'homes without hands.' Some, however, there are which are as fully entitled to be ranked among the true burrowers, as any creature which we have mentioned, digging a regular burrow in the earth, residing in their subterranean home, issuing forth to procure food, and retiring to it when alarmed. These are the creatures so widely famous as LAND CRABS (*Gecarcinus*), respecting which so many wonderful tales are told, some true, some false, and many exaggerated. The Land Crabs are found in various parts of the world, and are notable for very similar habits. They all burrow in the ground, run with very great speed, bite with marvellous severity, and associate in considerable numbers. As a general fact, they are considered as great dainties, and when properly prepared, may be ranked among the standing luxuries of their country.

As the VIOLET LAND CRAB of Jamaica (*Gecarcinus ruricola*), is the most familiar of these creatures, we will take it as our first example of the burrowing crustacea. This species, which is sometimes called the Black Crab, and sometimes the Toulourou, is exceedingly variable in its colouring, sometimes black, sometimes blue, and sometimes spotted. Whatever may be the colour, some tinge of blue is always to be found, so that the name of Violet Crab is the most appropriate of the three. Wherever the Land Crab makes its home, the ground is filled with its burrows, which are as thickly sown as those of a rabbit warren, and within these habitations the crabs remain for the greater part of the day, coming out at night to feed, but being always ready to scuttle back at the least alarm.

Although these warrens are seldom less than a mile from the sea, and are often made at a distance of two or even three miles, the Land Crabs are obliged to travel to the shore for the purpose of depositing their eggs, which are attached to the lower surface of the abdomen, and are washed off by the surf. Large numbers of the crabs may be seen upon their journey, which they prosecute so eagerly that they suffer no opposition to deter them from their purpose. This custom has probably given rise to the greatly exaggerated tales that have been narrated respecting these crabs, and their custom of scaling perpendicular walls rather than turn aside from the direct line of their route.

Twice in the year the Land Crabs become very fat and heavy, and are then in the best condition for the table, their flesh being peculiarly rich and loaded with fat. No one seems to be tired of the Land Crab, and new comers are apt to indulge in the novel dainty to such an extent that their internal economy is sadly deranged for some little time after the banquet.

About the month of August, the Land Crab is obliged to

cast its shell, and for that purpose retires to the burrow, which has been well stocked with grass, leaves, and similar materials. It then closes the entrance, and remains hidden until it has thrown off its old shell, and indued its new suit, which is then very soft, being little but a membranous skin, traversed by multitudinous vessels. At this time the crab is thought to be in the best condition for the table. Calcareous matter is rapidly deposited upon the membrane, and in process of time the new shell becomes even harder and stronger than that which has been rejected.

Many species of Land Crab are known, some of which possess rather curious habits. The FIGHTING CRAB (*Gelasimus bellator*), is a good example of them. This species possesses one very large and one very little claw, so that it looks as if a small man were gifted with one arm of Hercules and the other of Tom Thumb. As it runs along, with the wonderful speed which belongs to all its kin, it holds the large claw in the air, and nods it continually, as beckoning to its pursuer. While so engaged it has so absurd an aspect that it has earned the generic title of Gelasimus, *i.e.* laughable. As may be conjectured from its popular name, it is a very combative species, holding its fighting claw across its body, just as an accomplished boxer holds his arm, and biting with equal quickness and force. It is also a burrower, and lives in pairs, the female being within, and the male remaining on guard at the mouth of the hole, his great fighting claw across the entrance.

Another Land Crab, which has earned the generic title of Ocypode, or Swift-footed, and is popularly called the RACER, from its astonishing speed, is a native of Ceylon, where it exists in such numbers that it becomes a terrible nuisance to the residents. Having no respect for the improvements of civilisation, this crab persists in burrowing into the sandy roads, and is so industrious at its excavations, that a staff of labourers is constantly employed in filling up the burrows which these crabs have made. Were not this precaution taken, there would be many accidents to horsemen.

The mode of excavation employed by this creature is rather

peculiar. It 'burrows in the dry soil, making deep excavations, bringing up literally armfuls of sand, which, with a spring in the air, and employing its other limbs, it jerks far from its burrows, distributing it in a circle to the distance of many feet.'

THERE is a very remarkable burrowing crustacean, called the ROBBER CRAB (*Birgus latro*). This creature is of a strange, weird-like shape, difficult to explain, as it is unlike the form of most land-frequenting crustacea. The reader can, however, form some notion of its general form, by removing a common hermit crab from its residence, and laying it flat before him. The Robber Crab, however, does not live in a shell, and its abdomen is consequently defended by hard plates, instead of being soft and unprotected like that of the hermit crab, to which it is closely allied.

The Robber Crab inhabits the islands of the Indian ocean, and is one of those crustacea which are able to exist for a long time without visiting the water, the gills being kept moist by means of a reservoir on each side of the cephalothorax, in which the organs of respiration lie. Only once in twenty-four hours does this remarkable crab visit the ocean, and in all probability enters the water for the purpose of receiving the supply which preserves the gills in working order.

It is a quick walker, though not gifted with such marvellous speed as that which is the property of the racer and other land crabs, and is rather awkward in its gait, impeded probably by the enormous claws. While walking, it presents a curious aspect, being lifted nearly a foot above the ground on its two central pairs of legs, and if it be intercepted in its retreat, it brandishes its formidable weapons, clattering them loudly, and always keeping its face towards the enemy. Some travellers aver that it is capable of climbing up the stems of the palm-trees, in order to get at the fruit, and this assertion has lately been corroborated by the experience of competent observers.

The food of the Robber Crab is of a very peculiar nature, consisting chiefly, if not entirely, of the cocoa-nut. Most of my readers have seen this enormous fruit as it appears when taken from the tree, surrounded with a thick massy envelope of fibrous substance, which, when stripped from the nut itself, is employed for many useful purposes. How the creature is to feed on the kernel seems quite a mystery; and, *primâ facie*, for a crab to extract the cocoa-nut from its envelope, to pierce the thick and stubborn shell, and to feed upon the enclosed kernel, seems an utterly impossible task. Indeed, had not the feat been watched by credible witnesses, no one who was acquainted with the habits and powers of the crustacea would have credited such an assertion. Yet Mr. Darwin, Messrs. Tyerman and Bennett, and other observant men, have watched the habits of the creature, and all agree in their accounts.

According to Mr. Darwin, the crab seizes upon the fallen cocoa-nuts, and with its enormous pincers tears away the outer covering, reducing it to a mass of ravelled threads. This substance is carried by the crabs into their holes, for the purpose of forming a bed whereon they can rest when they change their shells, and the Malays are in the habit of robbing the burrows of these stored fibres, which are ready picked for them, and which they use as 'junk,' *i.e.* a rough kind of oakum, which is employed for caulking the seams of vessels, making mats, and similar purposes. When the crab has freed the nut from the husk, it introduces the small end of a claw into one of the little holes which are found at one end of the cocoa-nut, and by turning the claw backwards and forwards, as if it were a bradawl, the crab contrives to scoop out the soft substance of the nut.

PASSING by many other species of crustacea which burrow in the earth, or mud, or sand, we come to a very remarkable being, which makes its habitation in solid wood. This is the WOOD-BORING SHRIMP (*Chelura terebrans*), one of the sessile-eyed crustacea, nearly related to the well-known sand-hopper, which is so plentiful on our coasts.

Although very small, it is terribly destructive, and does no small damage to wooden piles driven into the bed of the sea. It is furnished with a peculiar rasping instrument, by means of which it is enabled to scrape away the wood and form a little burrow, in which it resides, and which supplies it with nourishment as with a residence. The tunnels which it makes are mostly driven in an oblique direction; so that when a large number of these creatures have been at work upon a piece of timber, the effect of their united labours is to loosen a flake of variable dimensions. As long as the weather is calm, the loosened flake keeps its position; but no sooner does a tempest arise, than the flake is washed away, and a new surface is exposed to the action of the Chelura.

When the Chelura is placed on dry land, it is able to leap nearly as well as the sand-hopper, and performs the feat in a similar manner.

THIS is not the only wood-boring crustacean with which our coasts are pestered; for the GRIBBLE (*Limnoria terebrans*) makes deeper tunnels than the preceding creature, though it is not so rapidly destructive, owing to the direction of its burrows, which are driven straight into the wood, and do not cause it to flake off so quickly as in the case when the Chelura excavates it. Still, it works very great harm to the submerged timber, boring to a depth of two inches, and nearly always tunnelling in a straight line, unless forced to deviate by a nail, a knot, or similar obstacle. The Gribble is a very tiny creature, hardly larger than a grain of rice, and yet, by dint of swarming numbers, it is able to consume the wooden piles on which certain piers and jetties are supported; and in the short space of three years these destructive crustacea have been known to eat away a thick fir plank, and to reduce it to a mere honeycomb. Sometimes these two wood-boring shrimps attack the same piece of wood, and, in such cases, the mischief which they perpetrate is almost incredible, considering their small dimensions and the nature of the substance into which they bore. The common fresh-water shrimp, so plentiful in our brooks and rivulets, is closely allied to the Gribble, and will convey a very good idea of its appearance. In some parts of our coasts the ravages of these animals are so destructive, that the substitution of iron or stone for wood has become a necessity.

CHAPTER V.

BURROWING MOLLUSCS

The BORING SNAIL of the Bois des Roches—Opinions as to the method of burrowing—Shape of the tunnels—Solitary habits of the Snail—The GAPER SHELL—The LIMPET—The PIDDOCK, its habits and appearance—Structure of the Shell, and its probable use—Method of burrowing—The DATE SHELL—Its extraordinary powers of tunnelling—The WOOD-BORER and its habits—The RAZOR SHELL—Its localities and mode of life—The SHIPWORM—Its appearance when young and adult—Its curious development—Its ravages—Its value to engineers.

ILL fitted as the Molluscs seem to be for the task of burrowing, there are several species which are able not only to make their way through soft mud, or into the sandy bed of the sea, but to bore deep permanent tunnels into stone or wood. Even the hard limestone and sound heart-of-oak timber cannot defy these indefatigable labourers, and, as the sailor or the dweller on the coast knows full well, the rocks and the timber are often found reduced to a mere honey-combed or spongy texture by the innumerable burrows of these molluscs.

THERE is now before me a piece of very hard calcareous rock, in which are bored several deep holes, large enough to admit a man's thumb, and remarkably smooth in the interior, the extremity being always rounded, Indeed, if a hole were made in a large lump of putty by putting the thumb into it and turning it until the sides of the hole became smooth, a very good imitation of these miniature tunnels would be produced. This fragment of stone was taken from a little wood in Picardy, called Le Bois des Roches, on account of the rocky masses that protrude through its soil, and was brought to England by Mr. H. J. B. Hancock, who kindly presented it to me.

In the winter time, each of these holes is occupied by a specimen of the Helix saxicava, a small snail, closely resembling the common banded snail of our hedges (*Helix nemoralis*), and it is thought that the holes are excavated by the snail which inhabits them. Mr. Hancock, who re-opened in the columns of the *Field* newspaper a controversy respecting these snails, which was initiated in 1839, is of opinion that the snails really form the hole, and that they burrow at the average rate of half an inch per annum. The late Dean Buckland was of the same opinion. Other naturalists, however, think that the holes were originally excavated by pholades and other marine molluscs when the rocks in question formed part of the ocean bed, and that the snails merely inhabit the ready-formed holes. Mr. Pinkerton upholds this opinion, and states that at least three other species of helix possess similar habits, the garden and the banded snail being among the number.

I have compared the burrows of the mollusc, which we will call the Boring Snail, with those of the pholas and lithodomus, both of which will be presently described, and find that there is no resemblance in their forms, the shape and direction of the holes being evidently caused by an animal of no great length in proportion to its width. In my own specimen, every hole is contracted at irregular intervals, forming a succession of rounded hollows. If we return to our lump of putty, we may form the holes made by the thumb into a very good imitation of those in which the Boring Snail lives. After the thumb has been pushed into the putty and well twisted round, put in the fore-finger as far as the first joint and turn it round so as to make a rounded hollow. Push the finger into the hole as far as the second joint, and repeat the process. Now introduce the whole of the finger, enlarge the extremity of the hole and round it carefully, when there will be a very correct representation of the tunnel formed in the rock.

Granting that the snail really does form the burrow, we have still to discover the mode of working. Mr. Hancock says that it must do so by means of an acid secretion proceeding from the foot, which corrodes the rock and renders it easy to be washed away. If the snail be removed and placed on litmus paper, the ruddy violet colour which at once tinges the paper shows that there is acid of some kind, and if the paper be applied to the spot whence the snail has been taken, the same results follow. It is a remarkable fact that although the snail leaves the usual slimy marks of its progress when crawling in the summer time, no mucus is perceptible on the approach of winter. When the cold months come round, the Boring Snail leaves its food and attaches itself to the rock, remaining in the same spot until summer approaches. During this time, the portion of rock to which it clings is worked away, and the stone around the excavation is impregnated with a greasy matter which soon dries up after the admission of the atmosphere. In a letter to me, dated October 14th, 1863, Mr. Hancock remarks that the rock at Monte Pellegrino in Sicily, which is crystalline and hard as marble, is perforated by the same snail and in the same manner. I may here mention that the stone of the Bois des Roches is that of which the column at Boulogne is built, which has retained its sharpness of outline after exposure to wind and weather for nearly sixty years. It is therefore called *marbre Napoléon.* Mr. Hancock proceeds to say, 'The following are a few of the peculiarities which I have not mentioned in my letter in the *Field*:

'1st. There is no instance at Bois des Roches of a tunnel being formed on the horizontal surface of a rock, or on the sides facing the south and south-east. They are always on the sides facing the north or north-east.

'2nd. The snail forms no epiphragm.

[The 'epiphragm' is the barrier of hardened mucus with which snails mostly close the entrance of their shells. There are generally several epiphragms in each shell.]

'3rd. Though during the summer it leaves behind it the usual slimy mucus track; in the winter on returning to the rocks no track is perceptible except the corrosion of the rock by frequent passage. This would seem to point to a system of secreting organs for the acid, separate from that for the mucus.

'4th. Contrary to the usual habits of burrowing molluscs, who generally have a bed of muddy matter between their shells and the walls of their dwelling, the Helix saxicava keeps his tunnel perfectly clean and neat.

'5th. When the liquor alluded to as forming a fatty aureole round the tunnel penetrates into pre-existing clefts in the rock, it provokes the growth of a microscopic lichen, which also grows in the tunnels in places after the liquor has evaporated.

'6th. The tunnels of the Helix saxicava are always irregular, bearing no relation to the size or shape of the excavators, whereas, in other excavating molluscs, the shape of the hole always bears some relation to its occupant, and also the excavations are alike for all animals of the same species.'

There is an opinion that the gastric juice secreted in the stomach may be the means through which the tunnelling is conducted, and that instead of being employed as food within the body it is poured out upon the stone, so as to dissolve it, the softened substance being then removed by the foot. The Boring Snails do not congregate together during hibernation, as is the well-known custom of the garden species, but are always solitary. Sometimes two or even three are found in the same burrow, but then they are always at some distance from each other, and form supplementary tunnels of their own. In my own specimen there is a curious example of this peculiarity, where the snail has contrived to bore completely through the barrier that separates it from a neighbouring tunnel, and has made a hole as large as the keyhole of an ordinary writing-desk, and nearly of the same shape.

THERE are many marine boring molluscs, some of which excavate mud, others stone, and others timber. Of the mud-borers I have little to say, few of them possessing points worthy of notice. Perhaps the most noteworthy of these is the common GAPER SHELL (*Mya arenaria*), so called, because one end of the shell gapes widely, in order to permit the passage of a long and stout tube. In a specimen now before me, the tube is between three and four inches in length, and at the base is large enough to admit the thumb. As, however, it gradually tapers to the extremity, the aperture at the other end is scarcely capable of receiving the little finger. The walls of this tube are very thin and membranous, and it is more or less retractile, carrying within it the siphons through which the mollusc respires and takes nourishment.

The Gaper Shell inhabits sandy and muddy shores, and to an inexperienced eye is quite invisible. The shell itself, together with the actual body of the mollusc, is hidden deeply in the mud, seldom less than three inches, and generally eleven or twelve inches from its surface. In this position it would be unable to respire, were it not for the elongating tube, which projects through the mud into the water, and just permits the extremities of the siphons to show themselves, surrounded by the little radiating tentacles which betray them to the experienced shell-hunter. These tentacles or fringes are never seen in the dried specimens, and can only be partially preserved by plunging the animal into spirits of wine, glycerine, or other antiseptic liquid. The Gaper Shell is esteemed as an article of food by man, beast, and bird; for not only do human beings dig it up with tools, cook it, and eat it, but the wolves and the arctic fox scratch it out of the mud and eat it raw, and the various sea birds peck it out with their beaks, prize the shell open, and devour the contents.

THE well-known LIMPET is a kind of borer, though the holes which it excavates are of very trifling depth, and are probably made by the mechanical friction of the shell and foot against the rock, without any intention on the part of the animal. Those who have been accustomed to wander along the sea-shore must have noticed that the Limpet shells always sink more or less into the rocks on which they cling, and that in very old specimens which are covered with algæ and barnacles, the shells are often sunk fully half their depth into the solid rock. Grooves, too, of various depths may be seen in the same rock, showing the slow and tedious track which the Limpets have made over its surface, until they finally settled down into some convenient situation.

Our next example of the burrowing molluscs is the well-known Pholas, popularly called the Piddock (*Pholas dactylus*), the shells of which are extremely plentiful upon our coasts,

PHOLAS IN WOOD. LITHODOMUS. RAZOR SHELL. HOLES OF PHOLAS IN ROCK.

whether empty and thrown upon the beach, or still adhering to the living animal and deeply sunken in the rock. Almost in every part of our shores the Piddock is to be found wherever there is rock, and its dimensions and general appearance vary together with the locality. The chalk cliffs, which bound so many miles of our coast, are thickly studded with the burrows of the Piddock, which takes up its residence as high as the mid-water zone of the coast, and in some places is so plentiful, that the hand can scarcely be laid upon the rock without covering one or two of the holes.

The shell itself is extremely fragile, and of a rather soft texture, and its outer surface is covered with ridges, that sweep in the most graceful curves from the hinge to the edge, and bear some resemblance to the projections upon a file. Yet practical naturalists have proved that, by means of these tiny points and ridges, the Pholas is able to work its way into the rock; for not only can a similar hole be bored by using the shell as a bradawl is used to pierce wood, but the creature has actually been watched while in the act of insinuating itself into the chalk rock, a feat which was performed by gently turning the shell from right to left, and back again.

The Pholas burrows to a considerable depth, and if a piece of the rock be detached and broken to pieces by the hammer, it will be seen to be completely riddled with the perforations. Chalk-rock is mostly the richest in specimens, but even the hard limestone formations are penetrable by the fragile shell of the Pholas. It has been well remarked, that the size of the Pholas and the sharpness of its markings vary in inverse ratio to the hardness of the rock in which it burrows. From the softest sea-beds are taken the largest and most perfect shells, while those specimens which are obtained from the hard limestone rocks, are comparatively small, and the surfaces are rubbed nearly smooth. The very worst examples, however, are those which are found in gritty rocks, interspersed with pebbles. The shells that have burrowed into such substances are dwarfed, abraded, and often misshapen, and are valueless except to the physiologist.

PERHAPS the DATE SHELLS are even more powerful as burrowers than the molluscs which have just been mentioned. One species, the Fork-tailed Date Shell (*Lithodomus caudigera*), is able to bore into substances which the pholas cannot penetrate. It is truly a wonderful little shell. Some of the hardest stones and stoutest shells are found pierced by hundreds of these curious beings, which seem to have one prevailing instinct, namely, to bore their way through everything. Onwards, ever onwards, seems to be the law of their existence, and most thoroughly do they carry it out. They care little for obstacles, and if one of their own kind happens to cross their path, they quietly proceed with their work, and drive their tunnel completely through the body of their companion.

The precise method employed in excavation is at present unknown, for the shape of the shell, and the exactitude with which it fits the burrow, prove that the mollusc does not form its tunnel by means of the protuberances on the surface of the shell, and no other method of boring has at present been discovered.

THOSE who are fond of wandering on the sea-shore, will often have experienced tangible proofs of the existence of another burrowing mollusc, the RAZOR SHELL (*Solen ensis*).

In some parts of our coast it is impossible to walk on the mixed rock and sand, when the tide has receded, without noticing innumerable jets of water, which start from the ground without any perceptible cause, leap for a foot or so in the air and then disappear. On watching one of these miniature fountains, and looking at the exact spot whence it proceeds, two little round holes are generally seen in the sand, so close to each other as to resemble a keyhole, and large enough to receive an ordinary goosequill. If the finger be placed on the spot, or even if the foot descends heavily on the ground, the curious object vanishes far out of the reach of a probing finger. The jets are thrown up by the Solen, and the two little holes are the open extremities of the siphon, that wonderful instrument through which the creature obtains its nourishment.

THE reader will remember that the wood-bearing pholas always makes its burrow *across* the grain of the timber which it is commissioned to destroy. The SHIPWORM (*Teredo navalis*), on the contrary, always burrows *with* the grain, and never makes a transverse tunnel, unless turned from its course by some obstacle, such as a nail, or the burrow of another Teredo.

At first sight, few would perceive that the Shipworm belongs to the same class as the oyster and the snail, for it is long, slender, and worm-like in shape, from six to eight lines in diameter, and nearly a foot in length. One end is rather larger

SHIPWORM.

than the shaft, if we may use the term, and is furnished with a pair of curved and very narrow shell-valves, while the other is divided into a forked apparatus containing the siphon. The colour is greyish-white.

Such is the aspect of the Shipworm when adult, but in its early stages of existence it possesses a totally different form. When it first issues from the sheltering mantle of its parent, it is a little, round, lively object, covered with cilia, like a very minute hedgehog, and, by the continual movement of these appendages, passing rapidly through the water. It does not, however, retain this form for more than six and thirty hours, but undergoes a further process of development, and is then furnished with a distinct apparatus for swimming and crawling. It also possesses rudimentary eyes, and in that portion of the body which may be considered the head, there are organs of hearing resembling those of certain molluscs. When it has passed its full time in this stage of development, it fixes upon some favourable locality, and then undergoes its last change, which transforms it into the worm-like mollusc with which naturalists are so familiar.

The ravages committed by this creature are almost incredible. Wood of every description is devoured by the Shipworm, whose tunnels are frequently placed so closely together that the partition between them is not thicker than the paper on which this account is printed. As the Teredo bores, it lines the tunnel with a thin shell of calcareous matter, thus presenting a remarkable resemblance to the habits of the white ant. When the Teredos have taken entire possession of a piece of timber, they destroy it so completely, that if the shelly lining were removed from the wood, and each weighed separately, the mineral substance would equal the vegetable in weight.

The Shipworm has been the cause of numerous wrecks, for it silently and unsuspectedly reduces the plankings and timbers to such a state of fragility, that when struck by the side of a vessel, or even by an ordinary boat, large fragments will be broken off. I have now before me two specimens of 'worm-eaten' timber, one of which is so honey-combed by this destructive mollusc, that a rough grasp of the hand would easily crush it. Yet this fragment formed part of a pier on which might have depended a hundred lives, and which was so stealthily sapped by the submarine miners, that its unsound state was only discovered by an accident.

Another species of the same genus, *Teredo corniformis*, is remarkable for the locality in which it is found. This curious mollusc burrows into the husks of cocoa-nuts, and other thick woody fruits which may be found floating in the tropical seas. In consequence of the locality which it selects for its habitation, it cannot proceed in one direction for any great distance, and is obliged to make its burrows in a crooked form, which has earned for the creature the specific title of corniformis, or horn-shaped. Fossil woods are often found perforated with these burrows.

Destructive as it may be, the Shipworm will ever be an object of interest to Englishmen, inasmuch as its shell-lined burrow gave to Sir I. Brunel the idea which was afterwards so efficiently carried out in the Thames Tunnel. And, though from the alteration of surrounding circumstances, that wonderful monument of engineering skill has not until of late been so practically useful as was anticipated, it has proved of incalculable value as pioneer to the numerous railway tunnels of this and other countries.

CHAPTER VI.

BURROWING SPIDERS.

The SCORPION and its habits—The burrow of the Scorpion—How detected—Suicide among the Scorpions—Spiders and their burrows—The Atypus—The TARANTULA—Its ferocity and courage—The TRAP-DOOR SPIDER—Its tunnel and the lining thereof—Its appearance under the microscope—The 'Trap-door' itself, and its structure—Curious example of instinct—Activity of the Spider—Strength and obstinacy of the Trap-door Spider—An Australian Trap-door Spider.

AMONG the burrowers belonging to this order may be reckoned the well-known SCORPION, of which there are several species, resembling each other in their general appearance, their structure and their habits.

Scorpions are found in all the warmer portions of the globe, and under the tropics they may be said to swarm. They are, as a general rule, intolerant of light, creeping by day into every cranny that can shelter them from the unwelcome sunbeams, and often causing very great annoyance by this custom. Old travellers, who have learned by experience the habits of these creatures, do not retire to rest before they have carefully examined the bed and surrounding furniture, especially taking up the pillow, and seeing that no enemy has lodged within the folds of the bedding. The left hand is generally employed in lifting the clothes, while the right is armed with a boot-jack, or stout shoe, or some other convenient weapon, with which the Scorpion may be immolated to the just wrath of its discoverer before it can run off and hide itself afresh. Shoes, boots, and gloves are also favourite resorts of the Scorpion, which has caused many an inexperienced traveller to buy future caution at rather a dear rate.

Scorpions may be found everywhere, under every stone, and

in every crevice; and it not unfrequently happens that when a pedestrian is passing over a sandy bank, and happens to break away a portion of it with his feet, a great black scorpion comes tumbling down, rolling over and over among the sandy avalanche, disengaging itself with an angry snap of its claws and a savage whisk of its tail, and showing fight as if it expected immediate attack from some present enemy. In such cases, the Scorpion has been a true burrower, excavating a temporary dwelling in the sandy soil, and living therein during the day.

The burrows of the Scorpion can always be detected by the peculiar shape of the entrance, which is of a semilunar form, exactly fitting the outline of the animal which digs it. The shape of the aperture is not unlike that of the hole which is cut in the seats of wooden stools for the purpose of introducing the hand when they are lifted. Wherever the soil is suitable for their purpose, the Scorpions take every advantage of it, so that a great number of these venomous creatures may be found in a comparatively small space of ground. Captain Pasley, R.A., tells me that, while in India, he has often destroyed, in the space of an hour or so, more than forty Scorpions, which had dug their sandy burrows in his garden.

The semilunar shape of the entrance is an infallible indication of the inhabitant, and in order to find out whether the Scorpion is at home, a jug full of water is poured into the burrow. Scorpions detest water, and when they feel the stream pouring upon them, they issue from their holes in high dudgeon, their pincers preceding them and snapping wildly at the enemy. A fork or spade is then driven under the Scorpion, and its retreat being thus cut off, it is easily killed.

The same officer also mentioned, that he had repeatedly tried the experiment of surrounding the Scorpion with a ring of fire, and that it had invariably stung itself to death. The fiery circle was about fifteen inches in diameter, and composed of smouldering ashes. In every instance the Scorpion ran about for some minutes, trying to escape, and then deliberately bent its tail over its back, inserted the point of its sting between two of the segments of the body and speedily died. This experiment was repeated seven or eight times, and always with the same results, so that a further repetition would have been but a useless cruelty. The heat given out by the ashes was very trifling, and not equal to that which is caused by the noontide sun, a temperature which the Scorpion certainly does not like, but which it can endure without suffering much inconvenience. Generally, the Scorpion was dead in a few minutes after the wound was inflicted.

Many of the true spiders are among the burrowers, and, even in our own country, it is possible to see a sandy bank studded with their silk-lined tunnels.

There is such a bank that skirts a fir-wood near my house, the material being the loosest possible sandstone, scarcely hard enough in any place to resist a pinch between the fingers and thumb. About an inch or two above the soil, this sandstone is quite excavated by the spiders, and as the sandy sides of their tunnels would fall in were they not supported in some manner, every tunnel is carefully lined by a coating of tough webbing, very strong, very elastic, very porous, and yet not suffering one particle of sand to pass through its interstices. From the opening of each burrow a web is spread, looking very much like a casting net, with a hole through its middle. From this again, radiate a number of separate threads, which extend to a considerable distance from the entrance.

At the very bottom of its silken tunnel the living architect lies concealed, its sensitive feet resting on the web, so that it is enabled to perceive the approach of the smallest insect that crosses the spot which it has so elaborately fortified. It is curious to watch the various insects that are caught by different species of spiders. The common garden spider (*Epeira diadema*) enjoys the greatest variety of diet, and the water spider, of which we shall see something in a future page, is also capable of varying its food to a considerable extent. The Burrowing Spiders, however, of which there are several species, are much restricted in their diet, the chief food that is found in their webs consisting of small beetles and midges. These spiders belong to the family Agelenidæ.

ONE of the best, if not indeed the very best, examples of the British burrowing Arachnida is the remarkable species, *Atypus Sulzeri*, a creature which is so rare as to have received no English name. It is a small species, not half an inch in length, but it is a curiously-constructed being; and were it made on a larger scale, would be a really formidable species. Its jaws are long, sharply pointed, and remarkably stout at their bases—so stout, indeed, that, but for a remarkable adaptation of structure, it would not be able to see anything in front.

None of these spiders have a separate head, that part of the body and the thorax being fused together, and forming what is called by naturalists a 'cephalothorax,' *i.e.* a head-thorax. The same structure may be observed in the scorpion, and also in the common lobster, the shrimp, and other crustacea. The eyes, as in all spiders, are rather close together, and are placed upon the upper part of this cephalothorax; but so large are the bases of the jaws, that they rise far above the level of the cephalothorax: and if the eyes were placed in the ordinary manner would act like the 'blind' that is hung over the eyes of a bad-tempered bull. In order, however, to enable the spider to see objects in front, a sort of little turret rises from the cephalothorax, and on its summit are placed the eyes. Naturalists familiarly call this projection the 'watch-tower.'

This spider inhabits moist situations, and burrows into the banks, the direction of the burrow being at first horizontal and then sloping downwards. It is lined with a remarkably compact silken tube, beautifully white, and about half an inch in diameter. The upper part of the tube is rather larger than the lower, and projects from the earth, falling forward so as to form a flap, which protects the mouth of the burrow. Specimens of this remarkable spider have been obtained from several parts of England.

SEVERAL large spiders that live mostly upon the ground are confounded together under the general name of Tarantula. There is scarcely a part of the world where is not found some great Lycosa, or Wolf-spider, that is popularly called by the dreaded name of Tarantula, and feared lest its bite should produce the disease which was once so rife through Europe, and called Tarantismus. These are all more or less burrowers, and line their tunnels with a silken coating, so as to prevent the earth from falling in upon them. Some of them hunt about after prey, while others sit at the entrance of the den and wait for the approach of any passing insect, which they may seize and devour at their leisure in the safe retreat of the neighbouring burrow. In this tunnel their young are hatched, and, as soon as they can struggle themselves free from the egg, they clamber upon their mother's back, and there cling in heavy clusters, often hiding her shape by their numbers.

One species of spider that goes by the name of Tarantula is resident in Siberia, and hides in holes in the ground. The peasantry are greatly afraid of it, fancying that it will bite them, and that its bite will cause great injury. For their terrors there are really some grounds, inasmuch as the spider is a savage kind of creature; and if a knife be pushed into its den, it will rush out in a fury, and try to bite the blade. In all probability, however, it is not very venomous, for it is actually eaten by sheep as they graze.

Of all the burrowing spiders, however, none is so admirable an excavator as the Trap-door Spider of Jamaica, and none displays so much ingenuity in the arrangement of its burrow. Specimens of both the tunnel and the spider are now before me, and it is impossible to inspect them without admiration. When removed from the earth which surrounded it, the silken tube is seen to be double, the outer portion being thick, deeply stained of a ruddy brown, and separated into a great number of flakes, lying loosely upon each other. This outer covering is so thick, harsh, and crumpled, that it looks more like the rough bark of a tree than a spider's web, and its true nature would hardly be recognised even by the touch. The exterior of a common wasp's nest bears some resemblance to this part of the tube. Beneath this covering is an inner layer of a very different character. This is uniformly smooth to the eye, and of a silken softness to the touch. It is but slightly adherent in places to the outer tube, and can be separated without any difficulty and without injuring the one or the other.

The texture of the interior surface is quite unlike that of the

NEST OF TRAP-DOOR SPIDER.
(From a Specimen in the collection of Lieut.-Col. C. J. Cox.)

inner or outer tube, being nearly white and of a smoothness and consistency much resembling the rough and unsized paper on which continental books are usually printed. It is curiously stiff also, and is so formed that no one who saw it for the first time would be likely to guess at its real character. The microscope, however, reveals its true character at once. If the interior of the tube be submitted to a moderately low power, say from thirty to forty diameters, a curious sight is presented to the observer. The surface looks like very rough felt, covered with little prominences, and composed of threads twisted together without the least apparent order. The threads are very coarse, in comparison to ordinary spider-web, and seem to be stiff, as if covered with size or gum.

The entrance of the tube is guarded by the 'trap-door,' from which the spider takes its name. This is a flap of the same substance as the tube, circular in shape, so as to fit the orifice with perfect accuracy, and attached to the tube by a tolerably wide hinge, so that when it closes it does not fall to either side, but comes true and fair upon the opening which it defends. The inner surface of the trap-door is white and felt-like, and exactly resembles the interior of the tube, but its outer surface is covered with earth, taken from the soil in which the hole is dug. As the trap-door is flush with the surface of the ground, it is evident that, when it is closed, all traces of the burrow and its inhabitant are lost.

The spider is urged by a curious instinct to make its tunnel in some sloping spot, and to keep the hinge uppermost, so that when the inhabitant leaves its home, or retreats to the extremity of its burrow, the door closes of its own accord, and effectually conceals it. New-comers into the country which the Trap-door Spider inhabits are often surprised by seeing the ground open, a little lid lifted up, and a rather formidable spider peer about, as if to reconnoitre the position before leaving its fortress. At the least movement on the part of the spectator, back pops the spider, like the cuckoo on a clock, clapping its little door after it quite as smartly as the wooden bird, and in most cases succeeds in evading the search of the astonished observer, the soil being apparently unbroken, without a trace of the curious little door that had been so quickly shut.

The spider itself is an odd-looking creature, with rather short, but very powerful legs, and a most formidable pair of fangs. Altogether, it has so crustacean an aspect, that, in common with many other species, it is called by the French the Crab-spider. The length of the specimen now before me is about an inch and a quarter, exclusive of the legs.

It is nocturnal in its habits, and during the night it leaves its burrow and hunts for prey. Insects of various kinds fall victims to this spider, and at the bottom of its tunnel may be found the relics of its feast, often including the remains of tolerably large beetles. If, when it is within its home, the lid be lifted gently, the spider hastens to the entrance, hooks its hind legs to the silken lining of the lid, and the fore legs to the side of the tube, and resists with all its might.

Nothing short of actual violence will induce the Trap-door Spider to vacate the premises which it so courageously defends. It will permit the earth to be excavated around its burrow, and the whole nest to be removed, without deserting its home; and in this manner specimens have been removed and placed in positions where their proceedings could be watched. Some few months ago, several examples of the Trap-door Spider and its nest were to be seen in the reptile-room of the Zoological Gardens. Boldly as the spider guards its home, and energetic as it is while engaged in defence, it is no sooner removed from the burrow than it loses all its activity, remains fixed to the spot as if stupefied, or, at the best, walks languidly about without appearing to have any definite object in view.

Trap-door Spiders inhabit many parts of the world. In the British Museum is a curious specimen of a nest, which is furnished with two doors, one at each end. The door of one end is rather loosely and irregularly made, as is, indeed, the whole end of the nest; but, at the other extremity, the door is beautifully rounded, very smooth, and fitting with astonishing neatness into the aperture. This curious specimen was discovered in Albania, and presented by W. Wilson Saunders, Esq.

The gem of the collection, however, for accuracy and finish, is one that is the work of an Australian spider, and was found at Adelaide. Only the upper part of the tube is preserved, so as to show the valve which closes it; but no one who really takes an interest in natural history can pass this nest without pausing in admiration. The workmanship is wonderful, and the hole, with its cover, looks as if it had been made in clay, by means of the potter's wheel, so regular and true are its outlines. The hole itself is circular, but the door is semi-circular, the hinge extending across the middle of the aperture.

Two points in this door are specially worthy of notice, the one being that its edge, as well as that of the aperture, is bevelled off inwards, so that the accurate closure of the entrance is rendered a matter of absolute certainty. The second point is, that the outer surface of the door, together with the surrounding earth, is ingeniously covered with little projections, so that when the door is closed, the line which, on smooth ground, would have marked its presence is totally hidden. The shape of the door, too, is remarkable. Towards its hinge it is comparatively thin, but upon the edge it is very thick, solid, and heavy, so that its own weight is sufficient to keep it firmly closed. The 'hinge,' to which allusion has frequently been made, is not a separate piece of workmanship, but is a continuation of the silken tube which lines the tunnel. An exact imitation of its principle may be made by taking the cover of a book, and cutting it across from the inside, until all its substance except the cloth or leather is severed, and then bending the two portions back. The cloth or leather will then form a hinge precisely similar to that of the Trap-door Spider, the pasteboard taking the place of the earthen door.

CHAPTER VII.

BURROWING INSECTS.

HYMENOPTERA.

The SAÜBA ANT and its habitation—Use of the 'parasol' leaves—Mr. Bates' account of the insect—Enormous extent of the Dwelling—The DUSKY ANT—Its Strength and Perseverance—Man and Insect Contrasted—The BROWN ANT—Form of its Habitation—Regulation of Temperature—Necessity of Moisture—How the Ant constructs Ceilings—Mining Bees—The ANDRENA and its burrowing Powers—The SCOLIA, its Burrows and its Prey—The HUMBLE BEE—Its general Habits—Locality of its Dwelling—Development of the Young—The LAPIDARY BEE, its Colours, Disposition and Habits—The WASP—Its Food and Habitation—Materials and Architecture of the Nest—Disposition, Form, and Number of the Cells—Biography of a Queen Wasp, and History of her Nest.

THE burrowing INSECTS now come before our notice.

Were this work to be arranged according to the rigid systems of zoological schoolmen, the list of burrowing insects must have been headed by the beetles ; but, as the subject of the book is to describe the peculiar dwellings which are needful for the welfare of various animals, a different arrangement is necessary, so that a well-built home takes precedence over a well-developed animal. If we wish to select an order of insects which surpasses every other in the variety and excellence of their burrows, we turn at once to the Hymenoptera, a large and important group of insects, which includes the wasps, bees, ants, sawflies, ichneumons, and one or two other families. The greater number of these insects burrow in the ground; but others are remarkable for their wonderful powers of excavating the hardest wood, and of forming therein a series of beautifully made cells, for the protection of the future brood.

Turn we first to some exotic Ants which inhabit tropical America.

I HAVE felt considerable doubt whether the SAÜBA, or COUSHIE ANT (*Œcodoma cephalotes*), ought to be reckoned among the burrowers or the builders, inasmuch as it makes large excavations below the ground, and raises dome-like edifices on its surface. As, however, the burrows are very much larger than the buildings, I shall place it with the former class, reserving for the corresponding example of the building-insects the Termites, whose edifices are more important than their burrows. It must first be mentioned that, although this species has often been described as the Visiting Ant, it is in reality a distinct species, as will be seen in the course of a few pages.

The Saüba Ant is restricted to tropical America, where it exists in such vast profusion, that it oftentimes takes forcible possession of the land, and drives out the human inhabitants who have cultivated and planted it. Broad columns of these ants may be seen marching along, each individual carrying in its jaws a circular piece of leaf, about the size of a sixpence, which is held vertically by one of its edges. In the British Museum there is a specimen of a Saüba Ant, with the leaf still grasped in its jaws, the ruling passion strong in death. From this curious habit the creature is sometimes called the Parasol Ant, and many persons have thought that the leaves are carried in order to protect the insect against the hot sunbeams. The real reason, however, has been discovered by Mr. H. W. Bates, who has studied with great care the habits of this remarkable insect, and has disentangled its history from many doubts and difficulties.

There are, as is usual with all ants, three distinct ranks —namely, the winged, the large-headed, or soldiers, as they are popularly called, and the ordinary workers. The large-headed individuals are sub-divided into two classes, namely, the smooth-heads and rough-heads, the former wearing a polished, horny, translucent helmet, and the head of the latter being opaque and covered with hair. The large-headed ants do no ostensible work, all the labour falling to the lot of the workers. These creatures make raids upon the trees, always giving the preference to cultivated trees, such as the orange and the coffee, and cut away the leaves so fast that the growth is stopped, and the entire plant sometimes dies.

The use of the leaves is to thatch the domes of their curious edifices, and to prevent the loose earth from falling in. Some of these domes are of gigantic dimensions, measuring two feet in height and forty feet in diameter, the mightiest efforts of man appearing small and insignificant when the comparative dimensions of the builders are taken into consideration. Division of labour is carried out to a wonderful extent in these buildings, for the labourers which gather and fetch the leaves do not place them, but merely fling them down on the ground, and leave them to a relay of workers, who lay them in their proper order. As soon as they have been properly arranged, they are covered with little globules of earth, and in a very short time they are quite hidden by their earthy covering.

The functions performed by the large-headed ants are not very evident. Those with smooth fronts seem to do nothing but walk about. They do not fight like the soldier-termites, nor do they appear to exercise any rule over the workers. Moreover, they have no sting, and even when assaulted they scarcely ever resent the insult. The hairy-headed variety is still more enigmatical in its duties. 'If the top of a small, fresh hillock, one in which the thatching process is going on, be taken off, a broad cylindrical shaft is disclosed, at a depth of about two feet from the surface. If this be probed with a stick, which may be done to the extent of three or four feet without touching the bottom, a small number of colossal fellows will slowly begin to make their way up the smooth sides of the mine. Their heads are of the same size as the class No. 2, but the front is clothed with hairs instead of being polished, and they have in the middle of the forehead a twin ocellus, or simple eye, of quite different structure from the ordinary compound eyes on the sides of the head. This frontal eye is totally wanting in the other workers, and is not known in any other kind of ant. The apparition of these strange creatures from the enormous depths of the mine reminded me, when I first observed them, of the Cyclopes of

Homeric fable. They were not very pugnacious, as I feared they would be, and I had no difficulty in securing a few with my fingers. I never saw them under any other circumstances than those here related, and what their special functions may be I cannot divine.'

The subterranean galleries which these creatures form are of almost incredible extent—so vast, indeed, and so complicated, that they have never been fully investigated. A conjecture as to their size may be formed from the fact, that when sulphur smoke was blown into a nest, one of the outlets was detected at a distance of seventy yards. The Saüba has often done considerable damage to property, having pierced the embankment of a large reservoir, and let out all the water before the damage could be detected.

The winged class is composed of the perfect male and female, which take their departure from the nest in January and February. They are quite unlike the other workers and soldiers, being larger and darker, with rounder bodies and a more bee-like aspect. The female is a really large insect, measuring more than two inches in expanse of wing, and the body being equal in size to a hornet; but the male is much smaller, as is generally the custom with the insect race. Of the hosts which pour out of the nests, only a few individuals remain after a space of twelve hours, the nest having been devoured by birds and other insect-eating creatures. Those which survive address themselves to the founding of new colonies; and so prolific are these insects, that, in spite of the vast destruction wrought among the winged individuals, to whom alone the task of reproduction belongs, man often has to retire before them, and even his art cannot conquer them.

The Saüba is one of the very few ants that does not attack other creatures. The real DRIVER, or VISITING, or FORAGING ANT, of which there are several species, belongs to another genus, Eciton, which will be described among the building-insects.

MOST of the British ants are among the burrowers, hollowing out subterranean abodes of great extent, and constructing them upon some intricate plan, the principle of which is not very evident. The DUSKY ANT (*Formica fusca*) generally prefers banks with a southern aspect, in which it forms its elaborate dwelling. Like many other ants, it is somewhat of a builder as well as a miner, and can raise story upon story, as well as add them by excavation. This task is achieved by covering the former roof with a layer of fresh and moist clay, and converting it into a floor for the next story. Dry weather has the effect of retarding the ants in their labours, because they find a difficulty in procuring sufficient moisture wherewith to mix the clay.

The muscular power and the energy and endurance of the ant are truly wonderful; and if a human being, even if aided by tools, could perform such a day's work as was achieved by a single ant without them, he would be a wonder of the world. M. Huber had the curiosity and good sense to devote the whole of a rainy day to watching the proceedings of a single Dusky Ant. The insect began by scooping out a groove in the earth, about a quarter of an inch in depth, kneading the earth, which it removed into little pellets, and placing them on each side of the groove, so as to form a kind of wall. The interior of the groove was beautifully smooth and regular, and when completed it looked very like a railway cutting, and performed a similar office. After completing this task, it looked about and found that there was another opening in the nest to which a road must be made, and straightway set to work upon a second sunken path of a similar character, parallel to the first, and being separated from it merely by a wall of a third of an inch in height.

Compare the size of an ant with that of a man, and then see how vast are the powers of so small a creature. Taking all the calculations in round numbers, and very much to the disadvantage of the ant, we find that a single man, who would have achieved a similar work in a single day, must have acted as follows:—

He must have excavated two parallel trenches, each of seventy-two feet in length and four feet six inches in depth; he must have made bricks from the clay he dug out, and with them built a wall along each side of the trenches, from two to three feet in height and fourteen or fifteen inches in thickness; and lastly, he must have gone over the whole of his work again, and smoothed the interior until it was exactly true, straight, and level. All this work must also have been done without the least assistance, and the ground most be supposed to be filled with huge boulders, and covered with tree trunks, broken logs, and other impediments.

The most admirable subterranean architecture is perhaps that of the BROWN ANT (*Formica brunnea*), a species which is not very commonly known in this country, and is probably confined to certain localities. Its habitation and the mode of its construction have been carefully noted by M. Huber.

This ant works mostly at night, and during light, misty rain, the sunbeams being obnoxious, and heavy showers causing much inconvenience. The nest is a most complicated structure, composed of a series of stories, often reaching thirty or forty in number, and generally being built in a sloping direction. These stories are not composed of regular cells, like those of the bee, wasp, and hornet, but of chambers and galleries of very irregular form and dimensions, beautifully smoothed in the interior, and about one-fifth of an inch in height. The walls are about the twenty-fourth of an inch in thickness. The object of so many stories is to be able to regulate the heat and moisture of their establishments. If, for example, the sun is not very powerful, and the instinct of the little insects tells them that more heat is required in order to hatch the pupæ which are undergoing their metamorphosis, they take up the white burdens and carry them into the upper chambers, where the heat is greater than below.

Again, if there should be a heavy rain, which floods all the lower stories, nothing is easier for the inhabitants than to remove themselves and brood into the upper sets of chambers, where they will be secure from the inundation. On those days when the sun is peculiarly hot, the ants secure a more equable temperature, by removing the young brood to the central flats, if they can be so called, while they themselves can obtain the needful moisture from the lower parts of the nest, to which the sunbeams cannot penetrate. Were it not for this provision which they instinctively make, all building operations would be stopped during a drought, whereas, by descending to the cellars or crypts of the mansion, the ants can obtain sufficient clay for ordinary work.

In order to watch the ants closer, Huber constructed a kind of vivarium in which they could work, and supplied them with earth, sand, and other necessaries. As, in this artificial state of existence, the insects could not procure moisture from the depths of the earth, moisture from other sources was necessary. Whenever the insects had ceased to work, they could almost always be induced to renew their labours by dipping a stiff brush in water, and striking the hand upon it in such a manner that the water descended like very fine rain upon the earth. As soon as the formerly quiescent ants felt the grateful shower, they regained their activity, ran about with renewed energy, and set to work upon the soil, moulding it into little pellets, and testing each tiny ball with their antennæ before they applied it to the purposes for which it was made.

While some of the ants were engaged in this task, which must be considered as analogous to brickmaking as practised by mankind, others were scooping out shallow hollows in the clay floor, the little ridges that were left standing being the foundation of the new walls. On these were dabbed the earthen pellets, and adjusted by means of the mandibles or by pressure of the fore feet, thus receiving compactness and uniformity. The most difficult part of such a task is the formation of the ceiling, but the ants do not appear to be at all embarrassed by so formidable an undertaking, but can lay ceilings of two inches in diameter with perfect certainty. The method of constructing the ceiling is by moulding the clay pellets into each angle of the chamber and also to the top of the pillars. As fast as one row of pellets becomes dry, a second is added; and the insects perform this delicate duty with such accuracy, that although so many centres are employed, the parts always coincide in the proper spots. The peculiar kneading and biting to which the clay pellets are subjected makes them exceedingly tenacious, so that they adhere strongly on the slightest contact.

As is well known, the ants do not retain their wings for any lengthened period, and after these members have served the purpose for which they were intended, they are broken off by the insect by means of a transverse seam near the base. There

MYRMELEON. AMPULEX. SCOLIA.

are, however, many of the permanently winged hymenoptera, which possess very great powers of burrowing, and are able to excavate soil so hard that a knife can scarcely make its way through the solidly impacted mass of earth and stones.

Mining Bees, which belong to the genus Andrena, are admirable burrowers, and, in spite of their small size, drive their little tunnels into the earth with astonishing ease. I once came on a whole colony of the Andrena, in a peculiarly hard and stony path near Dieppe. The ground was full of little holes, from which the bees were continually issuing, and into which others were as continually passing; their bodies yellow with the pollen of the flowers which they had been rifling, and which was intended to serve as a provision for the future brood.

An ordinary pocket-knife could make no impression on the ground, mixed as it was with stones, trodden by daily traffic, and baked by the heat of summer, into a mass nearly as hard as brick, harder perhaps than the bricks that are employed for modern houses. I was obliged, therefore, to return to my room and fetch a great, rude, thick-bladed clasp-knife that was reserved for rough work, and with much labour succeeded in tracing several of the burrows. They were sunk, on an average, about eight inches into the ground, and near the end they took a sudden turn, and were ended by a rounded chamber, in which was almost invariably a ball of pollen about as large as a pea. No larva was found in any of the burrows. The whole of the labour falls upon the female, the fore-legs of the male being unable to dig, and the hind-legs unable to carry the pollen.

At the right-hand side of the illustration on page 80 may be seen a figure of a remarkable burrowing bee, called *Scolia flavifrons*, a native of Europe, but not as yet proved to be British. In common with other fossorial bees, this insect is carnivorous in its larval state, and is supplied by its mother with the creatures on which it feeds.

This particular insect stocks its nest with the grub or larva of a beetle, belonging to the genus Oryctes. At the bottom of the cell may be seen certain grubs, the smaller of which is the larva of the Scolia, and the larger that of the beetle. As may be seen from the illustration, the grub of the beetle is very much larger than that of the creature which feeds upon it. The species which is here represented is a large and remarkably striking one, the four conspicuous spots at once distinguishing it from any other insect. In the middle of the illustration another example of a bee-burrower is given, in order to show the manner in which the insect takes its prey into the nest. The technical name of this species is *Ampulex compressa*, and its nest is stocked with cockroaches, one of which is being dragged into the hole, wherein it will be shortly eaten by the inhabitant.

In the illustration on page 83 are shown the nests of two common species of British HUMBLE BEE.

Both these species are burrowers, and sometimes make their nests at a considerable depth beneath the surface. The common HUMBLE BEE (*Bombus terrestris*) generally makes its subterranean house in the side of some bank, and the nest is usually found at a depth of a foot or eighteen inches. Sometimes, however, in places where the soil is light and friable, the nest has been found at a very great depth from the surface, so that a perpendicular shaft of five feet in length has been required before the nest could be reached. In all probability the bee has been aided by the burrow of a field mouse, when the gallery has been of such a length.

The history of the nest is really a curious one.

At the end of autumn, nearly all the Humble Bees die. The males invariably perish, but one or two of the females survive, and pass the winter in a state of hibernation. They do not select the nest for this purpose, convenient though the locality may seem, but hide themselves away singly in sheltered spots, such as the eaves of thatched barns, hollow trees, haystacks, or old ruins. When the sunbeams of spring gain warmth and strength, the sleepers awaken from their torpor, and immediately search for a spot wherein the new home may be excavated.

These bees, which are the Methuselahs of their short-lived race, may be seen in any warm spring day, flying about in all directions, prowling over every spare yard of ground, and settling here and there, as if to test the quality of the soil. They are very jealous of observation at this time, and if they think that they are being watched, will take instant offence and fly off with a quick, eager sound, very different from the steady, monotonous hum with which they accompany their researches. To watch one of these insects in hopes of seeing her begin her

BOMBUS TERRESTRIS. BOMBUS LAPIDARIUS.

labours, is an endless task, for she will never dig an inch of soil as long as she sees any suspicious object, and will often make her way under a thick tuft of herbage, and remain quietly in the retired nook until she fancies that the danger has passed away.

When, however, she has suited herself with a locality, she scrapes away the ground quickly, and when she has dug to a sufficient depth, she scoops out a small cavity or chamber, and therein constructs her first nest. There are but few cells at the beginning of the year, and these contain the first workers, who are intended to assist in constructing the enlarged nest. The larvæ are large, fat, white, round-bodied creatures, with little horny heads, and their bodies always slightly curved. When they have completed their feeding, each spins for itself an oval cocoon of coarse silk, rather irregular in shape, very soft, tough, and thick in consistency.

Herein they remain until they have attained their perfect state, when they gnaw a round piece from one end of the cocoon, just as a chicken chips off the top of the egg, and emerge into the nest. They do not venture out into the air for several days, the thick hair with which they are covered being all matted together, their wings soft and crumpled, and their limbs scarcely able to bear them. Two or three days are generally passed in the nest, and not until having gained their full strength do they venture out into the wide world. None but worker bees are developed for the first part of the year, the females and males not making their appearance until the summer weather has set in.

As may be seen from the illustration, the cells of the Humble Bee are not arranged in regular rows, like those of the hive bee, but are set carelessly side by side, mostly fixed together in groups of greater or lesser dimensions. Now and then a very little group of two or three cells is found, and single cells are occasionally to be seen, detached from the general mass.

THE right-hand nest in the illustration is that of the Red-tipped Humble Bee of Shakspere, known as the LAPIDARY BEE (*Bombus lapidarius*), which derives its specific name from its habit of making its nest within heaps of stone. This beautiful insect is plentiful in most parts of England, and may be known by the bright orange-red hue which decorates the last three segments of the abdomen. The female and worker of this species are precisely alike, except in their size; the former, which is popularly called the queen bee, measuring nearly an inch from the head to the tip of the tail, while the worker is scarcely half that length. The male is very variable in colour, but is generally black, with thick yellowish hairs upon the face, the fore part of the thorax, and the first segment of the abdomen.

I have always found this species to be fiercer than the preceding, and have more than once been driven away from the neighbourhood of the nest by its rapid and incessant attacks. The sting with which this bee is armed is a very formidable weapon, and the poison which it conveys into the wound is extremely virulent, causing much pain, and leaving a dull, aching sensation for several days afterwards. These symptoms, however, vary according to the individual who is stung, and those which are mentioned are described according to personal experience.

Generally, the Lapidary Bee makes its nest in heaps of stone, sometimes choosing those hillocks of rough stones which are heaped on the sides of roads, awaiting the stone-breaker and his hammer. Sometimes the fallen *débris* of limestone rocks affords a residence for this bee, and, in many instances, it burrows into the ground, and there makes its nest, just like that of the common humble bee.

THERE is one well-known and very handsome insect, which is equally disliked by the bee-keeper, the gardener, and the grocer, as it annoys them greatly in their respective callings. This is the common Wasp (*Vespa vulgaris*), which is equally fond of honey, fruit, and sugar; and as it is armed with a potent weapon, is not merely a hateful marauder, but a formidable enemy. The gardener, however, is the least injured of the three, for the Wasp confers upon him some slight benefits, which counteract in some degree the inroads which it makes upon his treasures. It is true that the Wasp is very fond of ripe fruit, and that with an unfailing instinct it prefers the choicest fruits, exactly when they are in their best condition, gnawing holes in them, and spoiling them for the market. Still it is more of a predaceous than a vegetable-feeding insect, and kills so many flies that it relieves the gardener of other foes, which, in the end, would be more injurious than itself, inasmuch as their larvæ endanger not only the fruit but the very life of the plant. It is a strangely bold insect, and has recourse to singular methods of procuring food. In the farming department at Walton Hall, I have seen the pigs lying in the warm sunshine, the flies clustering thickly on their bodies, and the Wasps pouncing on the flies and carrying them off. It was a curious sight to watch the total indifference of the pigs, the busy clustering of the flies, with which the hide was absolutely blackened in some places, and then to see the yellow-bodied Wasp, just clear the wall, dart into the dark mass, and retreat again with a fly in its fatal grasp. On the average, one Wasp arrived every ten seconds, so that the pigsty must have been a well-known storehouse for these insects.

As is well known to every boy who has participated in the delight of taking a Wasp's nest, the habitation of the insect is mostly under ground, and is a marvel of ingenious industry. The shape is more or less globular, and the material of which it is composed is very much like coarse brown paper, though not so tough. If it be opened, a wonderful scene is disclosed; terrace upon terrace of hexagonal cells being arranged in regular rows, and enclosed in a shell of papery substance, some half-an-inch in thickness, which is evidently intended to prevent the earth from falling among the combs, as these cell-terraces are called.

We will now suppose ourselves to be present at the construction of the nest, and, Prospero-like, will see without being seen.

In the early days of spring, a Wasp issues from the place in which it has passed the winter, and anxiously surveys the country. She does not fly fast nor high, but passes slowly and carefully along, examining every earth-bank, and entering every crevice to which she comes. At last she finds a burrow made by a field mouse, or perhaps strikes upon the deserted tunnel of some large burrowing insect, enters it, stays a long while within, comes out again and fusses about outside, enters again, and seems to make up her mind. In fact, she is house-hunting, and all her movements are very like those of a careful matron selecting a new home.

Having thus settled upon a convenient spot, she proceeds to form a chamber, at some depth from the surface, breaking away the soil, and carrying it out piece by piece. When she has thus fashioned the chamber to her mind—for she has a mind—she flies off again, and makes her way to an old wooden fence which has stood for many years, and which, although not rotten, is perfectly seasoned. On this she settles, and, after running up and down for a little time, she fixes upon some spot, and begins to gnaw away the fibres, working with all her might, so eagerly engaged that even were we not invisible we might stand by and watch her proceedings. At last, she has gathered a little bundle of fibres, which she gnaws and works about until she reduces them to a kind of pulp, and then flies back to the burrow.

She now runs up the side of the chamber, and clings to its roof with the two last pairs of legs, while with the first pair, aided by her jaws, she fixes the woody pulp on the roof, kneading it until it forms a kind of little pillar. Another and another supply is brought, until this pillar which is pendent from the roof, like a *papier-mâché* stalactite, is completed. The Wasp now begins to form the comb, and at the end of the pillar she places three very shallow cells, of a cup-like shape, not hexagonal, as are the completed cells. In each of these little cups she deposits an egg, and then constructs a roof over them, made from the same material as the cells, but laid in a different manner, the length of the fibres being nearly at right angles to the centre of the proposed comb. More cells are then added, eggs are laid in them, and the roof extended over them.

The eggs that were laid in the first three cells are now

hatched, and have produced very tiny grubs, which are always hungry and require much attention. They grow rapidly, and, in proportion to their growth, the parent Wasp adds to the walls of their cells, so that the young grubs are suspended, with their heads downwards, as, indeed, is the custom with very many hymenopterous larvæ. The Wasp proceeds in her task, having all the cares of the nest upon her—the enlargement of the chamber, the building of the nest, the transport of materials, the deposition of the eggs, and the feeding of the ever-hungry grubs.

In due time, however, the oldest grubs cease to feed, spin a silken cover over their cells, and release their parent from further attendance upon them. In the cells they undergo the change to the perfect state, and, after they have passed a short season in retirement, they tear away the silken cover with their jaws, and come forth as perfect Wasps. As soon as they have gained strength to use their limbs, they take the heavy labours upon them, and the work goes merrily on, the mother Wasp having little to do but to deposit eggs in the cells as fast as they are made.

Before very long, the first cell-terrace is completely full, and more accommodation is needed. This is supplied in a very curious manner. Taking the junction point of these cells as the foundation, the Wasps construct several pendent pillars, exactly like the one which has already been described, and, by dint of adding cells to each, they all unite, and form a second terrace, below the first, the distance between them being just sufficiently large to permit the Wasps to cross each other. In this, as in the former terrace, all the mouths of the cell are downwards and their bases upwards, so that the bases of the second terrace form a floor on which the Wasps can walk while feeding the young contained in the first. A third, fourth, and fifth terrace are added in this manner, all alike, the cells being so small that the mother Wasp cannot even put her head into them

It will be seen, therefore, that, as insects never grow after they have assumed the perfect form, the Wasps which have been bred in these cells must be very much smaller than their parent. They are the worker wasps, or neuters, as they are sometimes called, whose entire life is devoted to labour, and who, in fact, are undeveloped females.

Now, however, a change takes place. The cells of which the next few terraces are composed are of very much larger dimensions than the others, and are intended for the purpose of hatching the grubs which will afterwards become perfect male and female wasps. It will be seen, therefore, that the workers are hatched in the earlier part of the year, and that the male and female do not make their appearance until the end of the season. The cell-terraces increase gradually in diameter until the fourth or fifth, when they usually decrease slightly, and in exact accordance with their enlargement the covering is extended over them.

At the end of the season, after successive bands of worker-wasps have passed through the cells, and the single generation of the males and females has come to maturity, the nest shows symptoms of dissolution. If there are any grubs still left in the comb, the workers at once change their behaviour. Instead of feeding and tending them with jealous care, instead of defending them at the risk of their own lives, they pull these helpless white things out of their cradles, carry them far out of the nest, and abandon them. It seems a cruelty, and so it is; but it is a cruel mercy, substituting a quick death by exposure, or, perchance, being eaten by birds, for a slow and lingering death by starvation within the nest. For the instinct of the workers tells them that their labour is over, and their course is run, and that in a short time they will all die of old age, so that the helpless nurslings in the cells would find no food, and must perish by starvation.

At last, the entire population deserts the nest, the workers die, and so do all the males, none of them surviving their brief wedlock for more than a few hours; and the majority of the females die also, some from exposure to cold, and others by a violent death. Those, however, that are fortunate enough to find a crevice in which they can lie dormant during the long months of winter, creep into that, and there remain until the following spring, when they emerge to be the queens and mothers of future colonies. It is a remarkable fact that the Wasp never passes the winter in the nest, convenient as that spot may seem, but always seeks some other place of refuge. The reader will now comprehend, that whenever a Wasp is seen in the spring tide, it is one of the females which have survived the winter, and is about to found a new colony.

CHAPTER VIII.

BURROWING BEETLES.

The TIGER BEETLE, and its habits—Beauty of the Insect, its Larva, and mode of life—Curious form of its Burrow—The SEXTON BEETLE, and its power of digging in the ground—The DOR BEETLE, and the substances into which it Burrows—Use of the Dor Beetle—The SCARABÆUS of Egypt and its wonderful Instincts—The Egg, the Grub, and the Cocoon—Cocoon in the British Museum—The MOLE CRICKET, its form and elaborate Dwelling—Its general Habits, and wide distribution—The FIELD CRICKET, and its Tunnels—The MAY-FLY, and its home—The ANT-LION, its form, food, and mode of life—The Pitfall and its structure—Mode of catching Prey—Perfect form of the Ant-Lion.

WE now come to the Burrowing Beetles, of which there are no few species.

FIRST among the British coleoptera comes the lovely TIGER BEETLE (*Cicindela campestris*), an insect which, though small, can challenge comparison with the most beautiful exotic specimens. It is the fiercest, handsomest, and most active of all the British coleoptera, using legs and wings with equal agility, running or flying with such speed that its form cannot be clearly defined, and settling on the ground or taking to wing with equal ease. As it darts through the air, the burnished surface of the abdomen flashes in the sunbeams as if a living gem had passed by, earning for its owner the popular title of Sparkler Beetle.

This insect is a mighty burrower, exhibiting, even in its larval condition, some of that fiery energy which actuates it when it has reached its perfect condition. Sandy banks are the chief resorts of the Tiger-Beetle, which in this country seems seldom or never to alight upon trees, restricting itself to bare and sandy soil. It even avoids those spots which are covered with grass and herbage, cares nothing for shade, and delights to settle upon banks with a southern aspect, and to run about upon soil that has been rendered so hot by the sun that the bare hand can hardly endure contact with its surface.

The larvæ are most remarkable beings. They are whitish in colour, and strangely moulded in form, the head being of enormous size, and of a horny consistency, and the eighth segment developed into a hump-like projection, carrying upon its upper surface a pair of bent hooks. The larva never is seen above the surface of the ground, and, indeed, never exhibits more than the smooth horny head and mandibles. It lives in perpendicular burrows, about a foot in depth, which it is able to traverse with great rapidity, and which are only just of sufficient diameter to permit the inhabitant to pass up and down.

It is a carnivorous being, feeding chiefly on insects, which it is able to capture, in spite of the apparent disadvantage under which it labours of being confined to one spot. The mode by which it obtains its daily food is as follows. Ascending to the upper portion of its burrow, it fixes itself firmly by means of its hooks, and then lays its jaws level with the soil. While in this attitude, it is almost invisible, and as soon as an insect passes by the ambushed larva, the sickle-like jaws grasp it, and it is dragged to the bottom of the tunnel, where it is devoured.

The burrow is made by the larva, and not by the parent, and is a work of some little time, the earth being loosened by means of the feet and jaws, and then carried to the surface on the flattened head.

OTHER beetles are in the habit of driving deep tunnels into the ground, wherein may be deposited the eggs which are destined to produce a fresh brood in the ensuing season. Our own country can boast of possessing many such beetles, but in the hotter parts of the world their number is quite wonderful.

Our first example will be the well-known SEXTON, or BURYING BEETLES. There are several species of Burying Beetles; but as their habits are very similar, they need not be separately described. Anyone who wishes to see them at work may do so by taking a dead mouse, bird, or piece of meat, and laying it on a soft spot of ground. I was about to add the frog to the number of objects for sepulture, but have omitted that creature because the porous nature of its skin causes it to dry up so rapidly, that the beetle will seldom take the trouble of burying it.

Sometimes, but very rarely, a pair of the beetles will come to the bait by daylight, their wide wings bearing them along with great speed; but in general they prefer night as the time to begin their work. If the bird be visited early in the morning, it will be no longer upon the surface of the ground, but will be half sunk below it, as though the earth had given way, just as a piece of dark cloth sinks into snow. If, however, the bird be removed, the cause of its gradual disappearance will be seen in the form of one or two beetles, sometimes black, and sometimes beautifully barred with orange. Then let the bird be replaced, and a trowel carefully introduced under it, so that the bird and beetles can be gently transferred to a vessel of earth and covered with a glass shade.

During the day, the beetles will mostly remain quiet; but in the evening they begin to be active. To dig a hole, and then to drag the bird into it, would be a task far beyond their powers, and they therefore employ another plan. They entirely burrow beneath the bird, emerging every now and then to scrape out the loose soil, walk round the bird, mount it as if to see how the work is proceeding, and then disappear afresh and renew their labours. Sometimes they dig rather too much on one side, and then they appear sadly puzzled, running round and round the bird, getting on it as if to press it down with their weight, pulling it this way and that way; and at last they do what they ought to have done at first, namely, disappear under the bird and scrape away the earth until the hole is large enough to allow the bird to sink into the required position.

THE beetle just mentioned conveys into its burrow the whole of the substance on which the grub is intended to feed; but those which we shall now examine select only a portion for that purpose. There is a very large tribe of beetles, of which the

British type is the common DOR BEETLE (*Geotrupes vulgaris*), sometimes called the Watchman, or Clock, whose heavy hum drones upon the ear in the evening, as the

'Beetle wheels his drowsy flight,'

and whose hard and notched head occasionally strikes against the face with a violence less agreeable to the man than to the insect, the latter being quite undisturbed by the shock.

Let us watch this beautiful insect, as it wheels through the air. Either by the development of the sense of smell, or by some sixth sense with which humanity is practically unacquainted, the beetle is made aware that the object of its search is at hand. The dull, monotonous buzz is immediately exchanged for a triumphant hum, the circling flight ceases, and the beetle darts through the air, with arrow-like rapidity, to the spot which it seeks. A few more circles, lessening at every round, and down it settles, on an object uninviting to Europeans, but in great favour with Hindoos, Kaffirs, and scarabæi, namely, a patch of cow-dung.

No sooner has it settled, than it dives downwards until it reaches the earth, and then bores a perpendicular hole, some eight inches in depth, and large enough to admit a man's finger. Having ascended to the surface, it carries a quantity of the cow-dung to the bottom of the burrow, deposits an egg, and ascends, repeating this process as long as its powers endure. There are several other British beetles which prepare the cradle for their offspring in a similar manner.

Merely to dig a hole, to place at the bottom of it the food which the young are intended to eat, and to fill it in with earth, is a process of great simplicity, and makes but few calls on the industry or ingenuity of the labourer. Some allied beetles there are, however, which feed their young on similar substances, and in like manner bury them in the earth, but which exercise extraordinary industry in the performance of the task. All the world has heard of the famous SCARABÆUS of the Egyptians (*Scarabæus sacer*), an insect which is found in many parts of the globe, and very much resembles the Dor beetle of our own country. This insect sets to work in a curiously systematic manner.

As soon as the sensitive organs of the Scarabæus announce to it that the desired substance is at hand, it proceeds to the spot, alights, and sets at once to work. First, it sinks a tolerably deep and perpendicular hole in the ground, and, having returned to the cow-dung, it separates a sufficient quantity for its purpose, lays an egg in it, and forms it into a rude ball. She, for the female insect is the worker, then begins a curious and laborious task. Seizing the ball between her hind feet, she begins to roll it about in the hot sunshine, not taking it direct to the shaft which she has sunk, but remaining near

COCOONS OF SCARABÆUS AND GOLIATH.

the spot. Should rain come on she ceases to roll, or should the ball be made just before sunset, she waits for the morning before recommencing her labour. The consequence of all this curious rolling about, is twofold; it accelerates the hatching of the enclosed egg by the exposure to the sunbeams, and it forms a thin, hard, clay-like crust round the soft material in which the egg reposes.

When the ball is sufficiently rolled, it is taken to the hole, dropped down and the earth filled in. The egg is very soon hatched, and from it proceeds a little white grub, which finds itself at once in the midst of food, and begins to eat vigorously. By the time it has devoured the whole of the contents of its cocoon—if the mere empty shell may be so called—it is ready for its change into the pupal form, and there lies in the earth until it again changes its form and becomes a perfect beetle.

Perhaps the most extraordinary of these cocoons is that which is represented in the illustration. This is made by one of the gigantic beetles of the tropics. The insect which made it has no English name, but is scientifically called *Goliathus Drurii.* This wonderful cocoon is as large as a swan's egg, and, as may be seen by reference to the illustration, has very thin walls in proportion to its size. It is strengthened by a remarkable belt, which runs around its centre, exactly like that of the bullet which is used for the two-grooved rifle. How the belt is formed is perfectly unknown, as is its use, unless the strengthening of the walls be its only object. I have carefully examined the cocoon itself, and specimens of the insect which made it, and can find nothing which affords the least clue to the difficulty.

There is no doubt as to the species of insect which made it, for the creature lies inside, a small portion of the ends of the elytra and part of one leg being visible through the fracture. The colour of the beetle is peculiarly beautiful, being rich dark chocolate, soft and deep as made of velvet, and upon the thorax and round the elytra are drawn broad streaks of creamy white. On account of the large dimensions of the cocoon, it has necessarily been reduced in size, but a common house-fly is introduced into the drawing, in order to show the comparative size of the cocoon and the insect.

Many of the Orthopterous insects are burrowers, either digging holes wherein they themselves reside, or preparing a subterranean habitation for their young.

The best-known and most important of these insects is the Mole Cricket (*Gryllotalpa vulgaris*), called in some places the CROAKER, or CHURR-WORM, on account of the peculiar sound which it produces. It is a truly wonderful insect, one of those beings, which for the sake of force, we may perhaps call the anomalies of nature, though, in fact, nature is perfectly harmonious, and can have no real anomalies. A cursory glance at the insect will at once point out its habits, for the general shape, as well as the strange development of the fore-limbs, and the peculiar formation of the first pair of feet, are so similar to the corresponding members of the mole that the identity of their pursuits is at once evident.

Like the mole, the insect passes nearly the whole of its life underground, digging out long passages by means of its spade-like limbs, and traversing them with some swiftness. Like the mole, it is fierce and quarrelsome, is even ready to fight with its kind, and if victorious, always tears to pieces its vanquished opponent. Like the mole, it is exceedingly voracious, and requires so much food, that if several of them be confined in the same cage and kept only for a short time without food, the strongest will fall upon the weakest, kill and devour them.

To procure the insect is no easy matter, for it always burrows to some considerable depth when the soil is so loose, and a labourer with a spade would find much difficulty in disinterring it. The recognised method of procuring these insects is, to mark their holes by day and to visit them at dusk, just when the insects, which are nocturnal in their habits, are beginning to be lively. A long and pliant grass-blade is then pushed into the hole, the end is grasped in the jaws of the offended inhabitant, and both grass-blade and Mole Cricket are drawn out together.

Just as the mole constructs a habitation distinct from its ordinary galleries, so does this insect form a chamber for domestic purposes apart from the tunnels which ramify in so many directions. Near the surface of the ground a really large chamber is constructed, measuring about three inches in diameter, and nearly one inch in height. It is made very neatly, and the walls are carefully smoothed. Within this chamber the

Mole Cricket deposits its eggs, which are generally from two to three hundred in number, and yellowish in colour. As the chamber lies so near the surface of the ground, the genial sunbeams are able to raise the temperature sufficient for the hatching of the eggs, which in due course of time produce the tiny young, little white creatures, very like the parent in shape, except that they have no wings. They do not attain the perfect state until the third year.

The black-bodied FIELD CRICKET (*Acheta campestris*) is also one of the burrowing Orthoptera, working tunnels of considerable depth, and living in them during the day. By night it comes out of its home and sits at the mouth, chirping away for hours together. The banks at the side of a road or lane are favourite resorts of the Field Cricket, and I have noticed the insect peculiarly plentiful in the roads and lanes between Ramsgate and Margate. Like the mole cricket, it is of a very combative nature, and may be drawn out of its tunnel by the simple process of pushing a grass-stem down the burrow. It is said that in France it is captured in rather a curious manner, an ant being tied to a thread and dropped into the hole. Being partly carnivorous, the cricket seizes the ant for the purpose of eating it, and is immediately dragged out of its house by the thread.

BEFORE leaving the earth-burrowers, it is necessary to mention the larva of the common May-fly, or Ephemera. Sometimes this larva hides itself under stones, but it often burrows under the muddy banks, and there constructs a very curious habitation. If a portion of the mud be carefully removed, it will be seen to be perforated by a series of holes, a few being nearly circular, but the greater part oval, the long diameter being horizontal, in order to suit the peculiar shape of the inhabitant.

These are the habitations of the Ephemera grub; and if the block of mud be laid open, so as to exhibit longitudinal sections of the holes, the spectator will perceive that each hole is double, the two tubes lying parallel to each other, and being in fact only one tube bent upon itself.

OUR last example of the earth-burrowing insects is a truly remarkable one. I allude to the celebrated insect known as the ANT-LION (*Myrmeleon formicarius*). In its mature state, it presents nothing worthy of remark, except, perhaps, the elegance of its form, and the delicacy of its wide gauzy wings, which much resemble those of a common Dragon-fly. But in its larval condition it is truly a wonderful being.

Though predaceous, and feeding chiefly on the most active insects, it is itself slow, and totally unable to chase them; and were it not furnished with some quality which serves it in the lieu of speed, it would soon die of hunger. The very look of the larva is enough to make the observer marvel as to its method of obtaining food. Thick, short, soft, and fleshy, the body is supported on six very feeble legs, of which the hinder pair only are employed for locomotion, and these can only drag it slowly backwards From the front of the head project a pair of long, slender, curved mandibles, which give the first intimation that the grub has anything formidable in its nature. These mandibles are curiously made, being deeply grooved throughout their length, and permitting the maxillæ, or inner pair of jaws, to play up and down them.

Inert and helpless as it may seem, this grub is a ruthless destroyer of the more active insects, and, moreover, seldom catches any but the most active. Choosing some sandy spot, where the soil is as far as possible free from stones, it begins to form the celebrated pitfalls by which it is enabled to entrap ants and other insects. Depressing the end of its abdomen, and crawling backwards in a circular direction, it traces a shallow trench, the circle varying from one to three inches in diameter. It then makes another round, starting just within the first circle, and so it proceeds, continually scooping up the sand with its head, and jerking it outside the limits of its trench. By continuing this process, and always tracing smaller and smaller circles, the grub at last completes a conical pit, and then buries itself in the sand, holding the mandibles widely extended.

Should an insect, an ant, for example, happen to pass near the pitfall, it will be sure to go and look into the cavity, partly out of the insatiable curiosity which distinguishes ants, cats, monkeys, and children, and partly out of a desire to obtain food. No sooner has the ant approached the margin of the pitfall, than the treacherous soil gives way, the poor insect goes tumbling and rolling down the yielding sides of the pit, and falls into the extended jaws that are waiting for it at the bottom. A smart bite kills the ant, the juices are extracted, and the empty carcase is jerked out of the pit, and the Ant-lion settles itself in readiness for another victim.

Sometimes, when a more powerful insect, such as a large wood-ant, or beetle, or perhaps a hunting spider, happens to fall into the pit, the Ant-lion does not obtain a meal on such easy terms. The victim has no idea of surrendering at discretion, but tries to scramble up the sides of the pit, and in its furious exertions, it brings down the sand in torrents, filling up the pit, making the slopes of the sides shallower, and so rendering its escape easy. Then there is a battle between the Ant-lion and its intended prey, the one bringing the sand into the pit and the other flinging it out again so as to restore the steepness of the sides, and to deepen the pit.

Sometimes a quantity of the sand flung by the Ant-lion happens to fall on the escaping victim, knocks it over, and enables the devourer to grasp it in the terrible jaws, which never open but to reject the dead and withered carcases; sometimes the insect is tired before the Ant-lion, and suffers itself to be captured; and sometimes, though very rarely, it succeeds in making its escape. In either case, the pitfall is quite out of shape, and instead of re-arranging it, the Ant-lion deserts it and makes another. Some writers have said that the Ant-lion flings the sand at its escaping prey with deliberate aim and intention. It does nothing of the kind, but only tosses the sand out as fast as its head can work, without aiming in any direction, or having any idea except to prevent the pit from being filled up.

Its earth-burrowing life does not cease until it assumes the perfect state. When it has passed its full time in the larval condition, and is about to change into a pupa, it spins a silken cocoon of a globular form, and therein remains until it is about to assume its perfect condition. The pupa then bites a hole through the side of the cocoon, and projects its body half out the aperture. The pupal skin then withers, bursts, and the perfect insect emerges. Scarcely has it taken the first few breaths of air, than its abdomen, which before was short, so as to be included within the cocoon, extends to nearly three times its original length, so as to resemble that of the dragon-fly; the curious antennæ unroll themselves, the wings shake out by degrees their beautiful folds, and in a short time the lovely insect is ready for flight. It is scarcely possible to imagine a more complete contrast than that which is exhibited by the larva and the perfect insect, and if the two are placed side by side, no one who was not aware of the circumstances would think that they are but two stages of the same insect.

If the reader will refer to the illustration on page 80, he will see a section of the pitfall, with the Ant-lion at the bottom, and a couple of ants falling into the trap. The Ant-lion belongs to the same order of insects as the dragon-fly, which it so much resembles.

CHAPTER IX.

WOOD-BORING INSECTS.

BEETLES—The SCOLYTUS and its ravages—Mode of forming the Tunnels—Curious instinct—Worm-eaten Furniture, its cause, and the best method of checking the Boring Insects—Ginger-borers—The WASP BEETLE, its shape, colours, and tunnelling powers—The MUSK BEETLE—Its beauty and fragrance—Difficulty of detecting the Musk Beetle—Its Burrows and their inmates—The RHAGIUM and its Cocoon—Wood-boring Bees—WILLOW BEE, its Tunnel and mode of making the Cells—Food of the Young—The POPPY BEE—The PITH-BORING BEES and their Habits—Structure of the cells and escape of the Young—Shell-nests of Bees—Wonderful adaptation to circumstances—How the Bee burrows—The HOOP-SHAVER-BEE—Gilbert White's description of its habits—The SIREX and its Burrow—Its ravages among fir-trees—CARPENTER BEE—Mode of making its burrow—The Goat Moth—Its unpleasant odour—Shape and colour of the larva—Its winter cocoons—Escape of the moth from the burrow—CLEAR-WINGS WOLF MOTH and HONEY-COMB MOTH.

WE now leave the earth-burrowers, and proceed to those insects which tunnel into wood and other substances.

BEETLES generally burrow while in their larval state, though there are some that do so when they have attained their perfect form, and are able to bore their way through wood or into the ground with wonderful ease.

Perhaps there is no wood-boring beetle which is known so well as the little insect which is called *Scolytus destructor.* I am not aware that it has a popular name that will distinguish it from other small beetles which bore into wood.

The accompanying illustration will probably call to the mind of the reader, the insect which now comes before our notice. If he should have examined the bark of certain trees, particularly that of the elm, he will often have seen that it is perforated with circular holes, very like those which are drilled into worm-eaten furniture, but of rather larger diameter. When I was a very little boy and first saw these holes, I thought that they had been made by shot, and in trying to pick out the shot with my knife, made the discovery that the holes were not due to firearms, but to insects.

If the bark be cut through, and then raised with the knife, the curious radiating system of tunnels will be exposed to view, and the observer will notice that, however these tunnels may vary

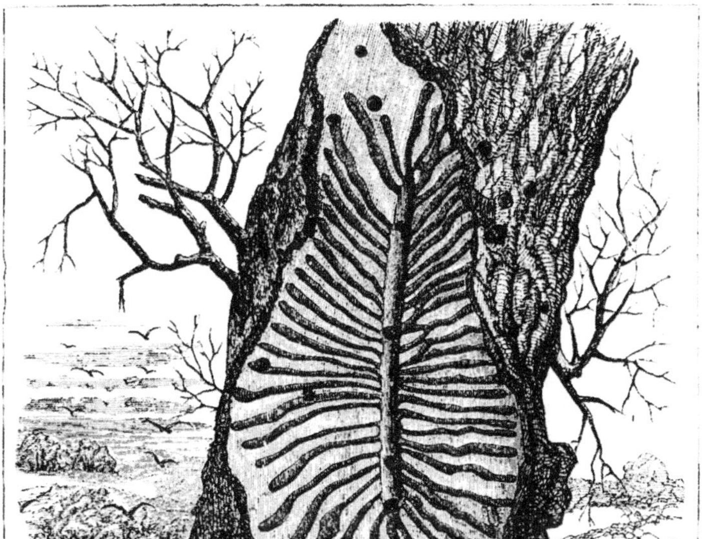

SCOLYTUS.

in size and direction, they all agree in these points; firstly, that they radiate nearly at right angles from a single cylindrical tunnel; and secondly, that they are very small at their base, and gradually increase to their termination. The cause of this formation is as follows:—

The mother insect enters the bark in search of food, and burrows deeply into the tree, sometimes boring into the substance of the wood itself, but generally cutting a tunnel between the wood and the bark. She then deposits her eggs regularly along the cylindrical tunnels, and in most cases retreats to the entrance, and there dies, her body forming a natural stopper. In due time the eggs are hatched, producing a number of very minute white grubs, which immediately begin to feed, the substance of the tree being the only diet of this insect in every stage of existence. Urged by a wonderful instinct, each grub arranges its body at a right angle with the burrow in which it was hatched, and so eats its way steadily outwards.

When the grubs have made some progress, the wisdom of this arrangement becomes evident. As they increase in size, the burrows necessarily increase with them, so that if they had all started parallel with each other, the tunnels would coalesce and the grubs be unable to procure their proper amount of food. As, however, the tunnels radiate like the spokes of a wheel, they very seldom interfere with each other, their radiation more than keeping pace with their increasing size. It will easily be seen by reference to the illustration, that if a number of these beetles attack a tree, the bark is gradually separated from the woody portion, and that, as in all exogenous trees the nourishment is derived from the bark, the tree must die as soon as the functions of the bark are suspended.

THE well-known 'worm-eaten' appearance of furniture is caused by certain beetles belonging to another family. As may be seen from the dimensions of the tunnels, the insects are very small, and their bodies are nearly cylindrical. The ravages which these beetles cause are fatal to all who happen to possess old furniture, but Mr. Westwood mentions that one common species, *Ptilinus pectinicornis*, completely destroyed a new bedpost, in the short space of three years. There is but one known method of killing the insects which have already taken possession, and of preventing others from following their example, namely, by injecting a solution of corrosive sublimate into the holes, and then treating the whole of the surface with the same poisonous liquid. I need perhaps scarcely mention, that insects which are popularly called Death-watches, belong to this family Not only do furniture and timber suffer from the attacks of the Ptilinus, but articles of dress and food are also injured by them. Specimens of natural history are often spoiled by the holes which are drilled through them by the beetles; and stationers sometimes suffer from the voracious insects, which bore holes through their wafers, fix them together, and there undergo their transformations within them. One species is obnoxious to wholesale druggists, on account of the damage which it does to the ginger. In some cases, half the ginger is drilled with holes, and rendered quite unsaleable. It is not, however, lost entirely, because it is reserved for the mill, and is then sold as ground ginger, the insects and their grubs being reduced to powder together with the ginger which they have not consumed. Such specimens are of course not exhibited to the general gaze, as the public would be very cautious of purchasing ground ginger if they knew what it contained. In the British Museum, however, may be seen several pieces of ginger completely eaten away by the beetle, and numerous examples of the insect itself are placed in the same tray.

THERE is a large group of beetles, which, in consequence of their extremely long antennæ, are called by the name of Longicornes. We have several examples in our own country, some of them being remarkable for the beauty of their colours, as well as for the elegance of their forms. The common WASP BEETLE (*Clytus arietis*) is a very good example of the longicorn beetles. It may be seen upon the hedges, gently slipping in and out with a curiously fussy movement, that very much resembles the restless gestures of the insect from which it takes its name. Its slender shape and yellow-striped body are indeed so wasp-like, that many persons are afraid to touch one of these beetles lest they should be stung.

The early life of the Wasp Beetle is spent entirely in darkness, the grubs burrowing into wood, and therein undergoing their transformations. They are curious little beings, white, roundish, but flattened; the rings of which the body is made are deeply marked, the segments nearest the head are much larger than those which compose the abdomen, and the head itself is small, but armed with a pair of jaws that remind the observer of wire nippers, so sharp are their edges, and so stout is their make. Old posts and rails are favourite localities with this beetle, and the grubs can almost always be obtained where timber has been left for any length of time in the open air.

ANOTHER well-known boring-beetle, is the large and beautiful insect which is popularly called the MUSK BEETLE (*Cerambyx moschatus*). Nearly an inch in length, with long and gracefully-curved antennæ, and slender and elegant in shape, it would always command attention, even if it were not possessed of two remarkable characteristics, colour and perfume.

To the naked eye, and in an ordinary light, the colour of this beetle is simply green, very much like that of the malachite. But, when the sun shines upon its elytra, some indications of its true beauty present themselves, not to be fully realised without the aid of the microscope and careful illumination. If a part of an elytron be taken from a Musk Beetle, placed under a half-inch object glass, and viewed through a good binocular microscope, by means of concentrated light, the true glories of this magnificent insect become visible. The general colour is green, but few can describe the countless shades of green, gold, and azure, that are brought out by the microscope, and no pencil can hope to give more than a faint and dull idea of the wonderful object. Neither do its beauties end with its colours, for the whole structure of the insect is full of wonders, and from the compound eyes to the brush-soled feet, it affords a series of objects to the microscopist, which will keep him employed for many an hour.

The odour which it exudes is extremely powerful; so strong. indeed, that I have often been attracted by the well-known perfume as I walked along a tree-fringed wood, and, after a little search, discovered the insect. It is no easy matter to find the Musk Beetle, even when it is close at hand, for its slender body lies so neatly along the twigs, and its green colour harmonizes so well with the leaves, that a novice will seldom distinguish the insect. A practised eye, however, looks out for the antennæ, and is at once attracted by their waving grace.

The larva of the Musk Beetle is a mighty borer, making holes into which an ordinary drawing-pencil could be passed. Old and decaying willow-trees are its favourite resort, and in some places the willows are positively riddled with the burrows. If such a tree be sawn open longitudinally, a curious scene is presented to the spectator. In some spots, the interior is hollowed out by nearly parallel burrows, until it looks as if it had been tunnelled by the shipworm, while sections are made of burrows that turn suddenly aside, or gradually diverge towards the yet uneaten parts of the timber. In some of the holes will be found the long white grubs, in others the pupa may be seen lying quiescent, while a perfect beetle or two may possibly be discovered near the entrance of the holes. Nor are the Musk Beetles the only tenants of the tree, for there is generally an assemblage of woodlice, centipedes, and other dark-loving creatures, which have crawled into the deserted holes, and taken up their abode within the tree.

We now come to the wood-boring bees, the name of which is legion, and a few examples of which will be now described.

The first is the Rose-cutter Bee (*Megachile Willoughbiella*), or Willow Bee, as it is often called, because its burrows are so frequently made in decaying willow-trees. This species is very common in most parts of England, and is therefore a good example of the wood-boring bees. The method by which the nests are made is very curious. After the insect has bored a hole of suitable dimensions in some old tree, she sets off in search of materials for the cells, and mostly betakes herself to a rose-bush, or laburnum-tree. She then examines one leaf after another, and having fixed on one to her mind, she settles upon it, clinging to its edge with her feet, and then, using her feet as one leg of a pair of compasses, and her jaws as the other, she quickly cuts out a nearly semicircular piece of leaf. As she supports herself by clinging to the very piece of leaf which she cuts, she would fall to the ground, when the leaf was severed, did she not take the precaution of balancing on her wings for a few moments before making the last cut. As soon as the portion of leaf is severed, she flies away with it to her burrow, and then arranges it after a truly curious fashion.

Bending each leaf into a curved form, she presses them successively into the burrow, in such a manner that they fit into one another, and form a small thimble-shaped cell. At the bottom of the cell she places an egg and some bee-bread, this substance being composed of pollen mixed with honey, and then sets to work upon another cell; and in this manner she proceeds until she has made a series of cells, some two inches in length. When the cells are first made, the natural elasticity of the leaf renders them firm, and as they become dry and stiff in a few days, they are then so strong that they can be removed from the burrow, and handled without breaking.

There is another bee allied to this genus, that employs the petals of the scarlet poppy for this purpose, but unfortunately it is not a native of England. Another species of burrowing bee, *Megachile centuncularis*, seems rather capricious in its choice of burrows, at one time making its tunnel into an old post or decaying tree, at another into the mortar of old walls, at another into the ground. It is extremely variable in size, sometimes barely exceeding a quarter of an inch in length, and sometimes reaching twice that size.

Very many species of pith-boring insects are known, most of them inhabiting the dry twigs of the bramble and garden rose. If at the cut end of a branch a round hole be found in the pith, the observer may be sure that a nest of some kind is within. Generally, on carefully laying the branch open, there appears a whole series of cells, one above the other, and in such a case, the cells which are farthest from the aperture always contain the larvæ of female insects, those nearest the entrance being the males.

Sometimes the nests which are found in the bramble contain the larvæ of *Osmia leucomelana*, a pretty little bee, scarcely more than a quarter of an inch in length, black in colour, with a very glossy abdomen, and a white, downy look about the legs.

Five or six cells are made in each branch, and the perfect insect appears about the month of June.

Several species select localities even more remarkable, and make their nests in the empty shells of snails. The common banded snail is a favourite with these bees, and in the British Museum may be seen a whole series of such nests. The number of cells necessarily varies with the size of the snail shell and the number of its whorls, but on the average four or five cells are found in each snail shell. The process of forming the cells is very simple. First, the bee deposits a quantity of pollen and honey, then she places an egg upon the pollen, and then she makes a partition with vegetable fibres torn by her teeth and kneaded firmly together. Lastly, the whole opening of the cell is closed by a wall formed of clay, tiny bits of stick, and small stones, and then the bee goes off in search of another shell. These shells may often be found under hedges, in moss, hidden by grass, and on examination the nests of bees will frequently be seen in them.

When the Osmia burrows into wood, she sets to work in a very deliberate manner. 'A bee,' writes Mr. F. Smith, 'is observed to alight on an upright post, or other wood suitable for its purposes. She commences the formation of her tunnel, not by excavating downwards, as she would be incommoded with the dust and rubbish which she removes; no, she work *upwards*, and so avoids such an inconvenience. When she has proceeded to the length required, she proceeds in a horizontal direction to the outside of the post, and then her operations are continued *downwards*. She excavates a cell near the bottom of the tube, a second and a third, and so on to the required number. The larvæ when full fed have their heads turned upwards. The bees which arrive at their perfect condition, or rather those which are first anxious to escape into day, are two or three in the upper cells—these are males; the females are usually ten or twelve days later. This is the history of every wood-boring bee which I have bred, and I have reared broods of nearly every species indigenous to this country.'

One of the wood-boring bees is especially worthy of notice, because some of its habits were remarked a century ago by Gilbert White, who did not know its name, but chronicled its method of obtaining padding for the nest. We will call it the Hoop-shaver (*Anthidium manicatum*). It is one of the summer insects, seldom appearing before the beginning of July, and is a rather stout-bodied insect, greyish black, with yellow lines along the sides of the abdomen. The last segment of the male is notable for its termination in five teeth. Its length is rather under half an inch, and it is a very remarkable fact that, contrary to general usage among insects, the male is larger than the female.

This bee seldom takes the trouble of making its own burrow, but takes advantage of the deserted tunnel of some other insect, such as the musk-beetle or the goat moth. When she has selected a fitting home, she enlarges it slightly at the end, and then goes in search of soft vegetable fibre wherewith to line it. 'There is a sort of wild bee frequenting the garden campion for the sake of its tomentum, which probably it turns to some purpose in the business of nidification. It is very pleasant to see with what address it strips off the pubes, running from the top to the bottom of a branch, and shaving it bare with the dexterity of a hoop-shaver. When it has got a vast bundle, almost as large as itself, it flies away, holding it secure between its chin and its fore-legs.'

After performing this part of her duty, she makes a number of cells, using the same material, together with some glutinous substance, placing an egg in each cell, and then leaves them. When the larvæ have obtained their full dimensions, they spin separate cocoons within the cells, and in the following summer the perfect insects make their appearance.

If the reader will visit any fir-wood, and look out for the dying and dead trees which are sure to be found in such places, he will probably see that many of them are pierced with round holes, large enough to admit an ordinary quill. These are the burrows of a splendid insect called *Sirex gigas* by entomologists.

Whether it has any popular name I do not know, but I have never been able to discover one, although I have shown specimens of the insect in many parts of England.

This is the more extraordinary, because it is really a splendid creature, nearly as large as a hornet, having wide wings, a bright yellow and black body, and a long firm ovipositor, so that from the head to the end of the ovipositor it measures an inch and three quarters in length. So unobservant, however, is the general public, that nine-tenths of those to whom I showed it declared that it was a wasp, and the remainder thought it to be a hornet.

The Sirex is a terrible destroyer of fir-wood, in some cases riddling a tree so completely with its tunnels that the timber is rendered useless. In a little fir-plantation about two miles from my house, there are a number of dead and dying trees, and almost every tree shows the ravages of this destructive insect. The absence of external holes is no proof that the Sirex has not attacked the tree, for they are only the doors through which the insect has escaped from the tree into the world.

The mode in which the Sirex carries on its operations is simple enough.

With the long and powerful ovipositor the mother insect introduces her eggs into the tree, and there leaves them to be hatched. As soon as it has burst from the eggs the young grub begins to burrow into the tree, and to traverse it in all directions, feeding upon the substance of the wood, and drilling holes of a tolerably regular form. Towards the end of its larval existence it works its way to the exterior of the trunk, and there awaits its final change, so that when it assumes its perfect form it has only to push itself out of the hole, and so finds itself in the wide world. The insects may often be seen on the trunks of the trees, clinging to the bark close to the hole out of which they have emerged.

THE Lepidoptera number among their ranks some of the most destructive wood-boring insects that inhabit this country.

There is, perhaps, no insect which makes so large or so ramified a burrow as the common GOAT MOTH (*Cossus ligniperda*). This insect is far more plentiful than is generally supposed, but as in its larval and pupal state it is deeply buried in some tree trunk, and in its perfect condition seldom ventures to fly by day, not one in a thousand is ever seen by the eye of man. This moth breeds in several trees, such as the willow, the oak, and the poplar, the first-mentioned tree seeming to be its chief favourite. Kent is one of the counties wherein this moth is found in greatest profusion, and in the fields round my house there is scarcely a willow of any size which has escaped the ravages of the Goat Moth caterpillar.

The larva of the Goat Moth derives its name from the very powerful and rank odour which it exhales, and which is thought to resemble that of the he-goat. This odour is not only strong but enduring, and for several years after the insect has vacated its burrow the disagreeable scent is plainly perceptible. I have now before me some specimens of the burrow of this creature, and although a very long time has evidently elapsed since the larvæ inhabited them, their odour is quite strong, and can be perceived at a distance of several feet. The pocket in which I placed them, after removing them from the tree, has never lost a rank reminiscence of its contents.

The larva is by no means a prepossessing creature, either to the eye or the nostrils, and though some persons believe that it was the famous Cossus, or tree-grub of the Romans, which was thought so great a delicacy by the ancients, I cannot believe that any palate could have attained so very artificial a condition as to endure this repulsive creature, much less to consider it as a dainty.

It grows with wonderful rapidity, being when it has reached its full size seventy-two thousand times heavier than when it was hatched; its segments are deeply marked, and in colour it is of a mahogany-red above, and yellowish below. The whole surface is smooth and polished, and, as may be presumed, considering the life which it leads, its muscular strength is enormous. Not only are the large and trenchant jaws extremely thick and strong, but the development of muscle is singularly great; and the head is of a wedge-like shape, so that the creature can force itself even through hard wood. It feeds entirely upon the substance of the tree in which it takes up its residence, and leaves in its tunnels a considerable amount of *débris*. As the creature increases in size, its tunnel increases in diameter; and it is an amusing task to cut up an old and soft-wooded tree, and follow the caterpillar through its manifold windings.

It lives for some three years in the larval condition, and during the winter it lies dormant in an ingeniously made cocoon, constructed from wood-chips and silken thread, a large store of which can be produced by this caterpillar. Some cocoons are now before me, which I took from a willow tree in Erith marshes Out of a great number of specimens I have selected four, in order to show the different dimensions of the cocoons. The largest is two inches and a quarter in length, and rather more than an inch in width. In shape it is nearly cylindrical, except at the ends, which are rounded. One of them is intact, but the other has a round hole through which the larva has emerged. It is composed of wood-chips of various sizes, looking like ordinary sawdust, which are loosely, though thickly, fastened upon a silken framework. Near one end of the cocoon the chips are very heavily massed, for what purpose seems doubtful. Rough, however, as is the exterior of the cocoon, the inside is quite smooth and soft, not unlike the interior of the tube made by the trapdoor spider.

The smallest cocoon is barely an inch in length, and is made of much smaller chips, fastened together so strongly that the cocoon retains its cylindrical form when handled, whereas the larger specimen is so loosely made that it collapses under the least pressure. The other two are intermediate in point of size, but precisely similar in point of construction. Besides them there is a specimen of the cocoon in which the creature undergoes its last change. This is of far stronger texture than either of the others, being quite hard, like *papier-maché*, and dark and polished within.

Generally, just before the moth emerges, the chrysalis works itself along, so that it partially projects from the hole, thus enabling the insect to escape at once into the outer world. In some instances, however, this is not the case, and in the present specimen the empty chrysalis shell may be seen, its shattered sides showing the manner in which the inclosed moth made its exit. The hole through which the moth emerged from the cocoon is of a wonderfully small size, considering the dimensions of the perfect insect, and its sides are very ragged and irregular. Like the other cocoons, it is strongly imbued with the characteristic odour, which has attached itself so strongly to my fingers that careful ablution will be needed before I shall venture to produce my hands in society.

SOME of the most elegant and curious British Lepidoptera are also among the most destructive.

The various species belonging to the remarkable family Egeriadæ, properly called Clear-wing Moths, are terrible enemies to the gardener, as well as to the landowner, their larvæ feeding upon the pith, and generally preferring the young wood to that of a more advanced growth. In some cases they live in the roots, and are quite as destructive as their relations who prefer the branches. All the Clear-wings are distinguished by the fact that the greater part of their wings is simply membraneous and transparent, without the beautiful feathery scales that are worn by the Lepidoptera as an order. Some of them resemble hornets, others are often mistaken for wasps, while several species are wonderfully like gnats, and as they fly about in the sunshine may readily be mistaken for these insects.

Of one of these insects, *Ægeria asiliformis*, known to collectors as the BREEZE-FLY CLEAR-WING, Mr. J. Rennie writes as follows: 'We observed above a dozen of them, during this summer, in the trunk of a poplar, one side of which had been stripped of its bark. It was this portion of the trunk which all the caterpillars selected for their final retreat, not one having been observed where the tree was covered with bark. The ingenuity of the little architect consisted in scraping the cell almost to the very surface of the wood, leaving only an exterior covering of unbroken wood, as thin as writing-paper. Previous, therefore, to the chrysalis making its way through this feeble barrier, it could not have been suspected that an insect was lodged under the smooth wood. We observed more than one of these insects in the act of breaking through this covering, within which there is besides a round moveable lid, of a sort of brown wax.'

The last-mentioned peculiarity is worthy of special notice, because it is not a general feature in the history of the Clear-wings. Just when they are about to change into the pupal form, they usually nibble a hole through the exterior of the branch, and then make a partial cocoon out of the *débris*, taking care to place themselves so that the head is towards the orifice. The abdominal segments of the chrysalis are furnished with points directed backwards, so that by alternately extending and contracting the abdomen, the creature is pushed onwards. When it is going to break out of its chrysalis case it uses these little points, and forces itself partially through the hole, thus allowing the perfect moth to issue at once into the world.

With two more species of lepidopteran burrowers, we must close our list, one of them boring into wood and the other into wax.

The first of these insects, *Tinea granella*, is sometimes called the WOLF MOTH. It is a very small insect, and is closely allied to the common clothes moth, so deservedly hated by fur-dealers, careful housewives, and keepers of museums. The larva of this insect feeds upon the corn, covering it at the same time with a tissue of silken threads. The most curious portion of the life of this insect is, that after the larva has finished eating the corn, it proceeds to the sides of the granary, and there burrows into the wood, making its holes so closely together that, if the timber had been taken out of the sea, the Gribble would have had the credit of the tunnels. Nothing seems to stop this little creature, and it bores through deal planks with perfect ease, making its way even through the knots without being checked either by the hardness of the wood, or the abundance of turpentine with which the knots in deal are saturated. This is the more astonishing, because turpentine is mostly fatal to insects, and a little spirit of turpentine in a box will effectually keep off all moths and beetles.

In these burrows the larvæ change into the pupal state, and there remain until the following summer, when they emerge in hosts, ready to deposit their eggs upon the corn, and raise up fresh armies of devourers. Another singular fact is, that after these caterpillars have lived for so long upon corn, their tastes should change so suddenly as to induce them to take to wood, and wood moreover which is never free from turpentine, however well it may be seasoned.

THE last of our burrowers is the HONEY-COMB MOTH, belonging to the genus Galleria. Two species of this genus are known in England, both of which are plentiful in this country.

These moths live in the comb of the hive bee, and when once they have succeeded in depositing their eggs, the combs are generally doomed. The envenomed stings of the bees are useless against these little pests, for though their bodies are soft they take care to conceal themselves in a stout silken tube, and their heads are hard, horny, and penetrable by no sting borne by bee. I once had a very complete case of honey-comb utterly destroyed by the Galleria moths, which drew their silken tubes through and through the combs, ate up even my beautiful royal cells, devoured all the bee-bread, and converted the carefully chosen specimens into an undistinguishable mass of dirty silk, *débris* and moths, both dead and living.

Although there are still in my list many names of burrowing insects which have not yet been described, it is necessary that we should take our leave of the burrowers, and proceed to the next chapter.

CHAPTER X.

PENSILE MAMMALIA.

THE HARVEST MOUSE—Its appearance—Reason for its name—Mouse nests—Home of the Harvest Mouse—A curious problem—Food of the Harvest Mouse, and its agility—The SQUIRREL—Its summer and winter 'cage'—Boldness of the Squirrel—Materials for the nest, and their arrangement.

THERE are not many mammalia which make pensile nests, and we are, therefore, the more pleased to find that one of the most interesting inhabits this country. This is the well-known HARVEST MOUSE (*Micromys minutus*), the smallest example of the mammalia in England, and nearly in the world.

This elegant little creature is so tiny that, when full-grown, it weighs scarcely more than the sixth of an ounce, whereas the ordinary mouse weighs almost an entire ounce. Its colour is a very warm brown above, almost amounting to chestnut, and below it is pure white, the line of demarcation being strongly defined. The colour is slightly variable in different lights, because each hair is red at the tip and brown at the base, and every movement of the animal naturally causes the two tints to be alternately visible and concealed.

It is called the Harvest Mouse, because it is usually found at harvest time, and in some parts of the country it is captured by hundreds, in barns and ricks. To the ricks it could never gain admission, provided they are built on proper staddles, were it not that it gets into the sheaves as they stand in the field, and is carried within them by the labourers. Other mice, however, are sometimes called by this name, although they have no fair title to it; but the genuine Harvest Mouse can always be distinguished by its very small size, and the bright ruddy hue of the back and the white of the abdomen. Moreover, the ears of the Harvest Mouse are shorter in proportion than those of the ordinary mouse, the head is larger and more slender, and the eyes are not so projecting, so that a very brief inspection will suffice to tell the observer whether he is looking at an adult Harvest Mouse, or a young specimen of any other species.

Mice always make very comfortable nests for their young, gathering together great quantities of wool, rags, paper, hair, moss, feathers, and similar substances, and rolling them into a ball-like mass, in the middle of which the young are placed. The Harvest Mouse, however, surpasses all its congeners in the beauty and elegance of its home, which is not only constructed with remarkable neatness, but is suspended above the ground in such a manner as to entitle it to the name of a true pensile nest. Generally, it is hung to several stout grass-stems; sometimes it is fastened to wheat-straws; and in one case, mentioned by Gilbert White, it was suspended from the head of a thistle.

It is a very beautiful structure, being made of very narrow grasses, and woven so carefully as to form a hollow globe, rather larger than a cricket-ball, and very nearly as round. How the little creature contrives to form so complicated an object as a hollow sphere with thin walls is still a problem. It is another problem how the young are placed in it, and another how they are fed. The walls are so thin that anything inside the nest can be easily seen from any part of the exterior; there is no opening whatever, and when the young are in the nest they are packed so tightly that their bodies press against the wall in every direction.

The position of the nest, which is always at some little height, presupposes a climbing power in the architect. All mice and rats are good climbers, being able to scramble up perpendicular walls, provided that their surfaces be rough, and even to lower themselves head downwards by clinging with the curved claws of their hind feet. It is also a noticeable fact, that the joint of the hind foot is so losely articulated that it can be turned nearly half round, and so permits great freedom of movement. The Harvest Mouse is even better constructed for climbing than the ordinary mouse, inasmuch as its long and flexible toes can grasp the grass-stem as firmly as a monkey's paw holds a bough, and the long, slender tail is also partially prehensile, aiding the animal greatly in sustaining itself, though it is not gifted with the sensitive mobility of the same organ in the spider monkey, or kinkajou.

As the food of the Harvest Mouse consists greatly of insects, flies being especial favourites, it is evident that great agility is needed. Its leap is remarkably swift, and its aim is as accurate as that of the swallow. Even in captivity, it has been known to take flies from the hand of its owner, and to leap along the wires of its cage as smartly as if it were trying to capture an insect that could escape.

The Harvest Mouse is tolerably prolific, and in the airy cradle may sometimes be seen as many as eight young mice, all packed together like herrings in a barrel.

THERE is another well-known British mammal which, at all events at one season of the year, may be classed among those creatures who build pensile nests. This is the common Squirrel, so plentiful in well-wooded districts, and so scarce where trees are few.

The Squirrel is an admirable nest-builder, though it cannot lay claim to the exquisite neatness which distinguishes the harvest mouse. As is well known, the Squirrel constructs two kinds of nests, or 'cages,' as they are popularly called, one being its winter home, wherein it can remain in a state of hibernation, and the other its summer residence. These two nests are as different as a town mansion and a shooting-box, the former being strong, thick-walled, sheltered, and warm, and the other light and airy. The winter cage is almost invariably placed in the fork of some tree, generally where two branches start from the trunk. It is well concealed by the boughs on which it rests, and which serve also as a shelter from the wind. The summer cage, on the contrary, is comparatively frail, and is placed nearly at the extremity of slender boughs, which bend with its weight, and cause the airy cradle to rock and dance with every gust of wind.

As if conscious of the impregnable situation which it has chosen, the Squirrel takes no pains to conceal the summer cage, but builds it so openly, that it can be seen from a considerable distance; whereas the winter home requires a practised eye to detect it. So confident is the animal in the strength of its position, that it can scarcely be induced to leave the nest, and will sit there in spite of shouts and stones, provided that the missiles do not actually strike the nest. A well-aimed stone will generally alarm the cunning little animal, and cause it to make one of its rapid rushes to the top of the tree. The materials of the Squirrel's cage are very similar to those of an ordinary bird's nest, consisting of twigs, leaves, moss, and other vegetable substances. Its structure is tolerably compact, though it will not endure rough handling without being injured.

In this aerial nest the young Squirrels are born, making their appearance in the middle of summer, and remaining with their mother until the following spring. There are generally three or four young; and though the nest appears to be so slight, it is capable of sustaining the united weight of young and parents. The Squirrel does not seem to make more nests than can be avoided, and, like many nest-builders, inhabits the same domicile year after year, until it is quite unfit for occupation. Should the nest be assailed while the young are still helpless, the mother takes them in her mouth one by one, leaps away with them, and deposits them in some place of safety. The shape of the summer nest is nearly spherical. The winter cage, however, is most irregular in form, being accommodated to the space between the boughs in which it is built, and is very thick and warm.

CHAPTER XI.

PENSILE BIRDS.

WEAVER BIRDS and their general habits—The MAHALI WEAVER BIRD—Shape of the nest—Singular defence—Remarkable nests of Weavers—Very curious contrivance—The GOLD-CAPPED WEAVER—Structure and situation of the nest—The TAILOR BIRD—Structure of the nest—The FAN-TAILED WARBLER—Singular method of fixing its nest.

ALTHOUGH the majority of nest-making birds may be called Weavers, there is one family to which the name is *par excellence* and with justice applied. These are the remarkable birds which are grouped together under the name of Ploceidæ, all being inhabitants of the hot portions of the old world, such as Asia and Africa.

For the most part, the Weaver Birds suspend their nests to the ends of twigs, small branches, drooping parasites, palm-leaves, or reeds, and many species always hang their nests over water, and at no very great height above its surface. The object of this curious locality is evidently that the eggs and young should be saved from the innumerable monkeys that swarm in the forests, and whose filching paws would rob many a poor bird of its young brood. As, however, the branches are very slender, the weight of the monkey, however small the animal may be, is more than sufficient to immerse the would-be thief in the water, and so to put a stop to his marauding propensities.

Snakes. too, also inveterate nest-robbers, some of them living almost exclusively on young birds and eggs, are effectually debarred from entering the nests, so that the parent birds need not trouble themselves about either foe. Although they may repose in perfect safety, undismayed by the approach of either snake or monkey, they never can see one of their enemies without scolding at it, screaming hoarsely, shooting close to its body, and, if possible, indulging in a passing peck.

All the pensile birds are remarkable for the eccentricity of shape and design which marks their nests; although they agree in one point, namely, that they dangle at the end of twigs, and dance about merrily at every breeze. Some of them are very long, others are very short; some have their entrance at the side, others from below, and others again, from near the top. Some are hung, hammock-like, from one twig to another; others are suspended to the extremity of the twig itself; while others, that are built in the palms, which have no true branches, and no twigs at all, are fastened to the extremities of the leaves. Some are made of various fibres, and others of the coarsest grass-straws: some are so loose in their texture, that the eggs can be plainly seen through them; while others are so strong and thick, that they almost look as if they were made by a professional thatcher.

A good example of the last-mentioned description of nest is the Mahali Weaver Bird of South Africa (*Pliopasser Mahali*). Although the architect is a small bird, measuring only six inches in total length, the nest which it makes is of considerable size, and is formed of substances so stout, that, when the edifice and the builder are compared together, the strength of the bird seems quite inadequate to the management of such materials.

The general shape of the nest is not unlike that of a Florence oil-flask, supposing the neck to be shortened and widened, the body to be lengthened, and the whole flask to be enlarged to treble its dimensions. Instead, however, of being smooth on the exterior, like the flask, it is intentionally made as rough as possible. The ends of all the grass-stalks, which are of very great thickness, project outwards, and point towards the mouth of the nest, which hangs downwards; so that they serve as eaves whereby the rain is thrown off the nest.

Perhaps the most singular-looking nest made by these birds is that of a rather small, yellow-coloured species (*Ploceus ocu-*

larius). This nest looks very like a chemist's retort, with the bulb upwards—or, to speak more familiarly, like a very large horse-pistol suspended by the butt. The substance of which it is made is a very narrow, stiff and elastic grass, scarcely larger than the ordinary twine used for tying up small parcels, and interwoven with a skill that seems far beyond the capabilities of a mere bird.

If the hand be carefully introduced up the neck of one of these nests, its admirable fitness for the nurture of the young birds is at once perceived. When merely viewed from the outside, the nest looks as if it would be a very unsafe cradle, and would permit the young birds to fall through the neck into the water. A section of the nest, however, shows that no habitation can be safer, and even the hand can detect the wonderfully ingenious manner in which the interior is constructed. Just where the neck is united to the bulb, a kind of wall or partition is made, about two inches in height, which runs completely across the bulb, and effectually prevents the young birds from falling into the neck.

Another of this group is the GOLD-CAPPED WEAVER BIRD, *Ploceus icterocephalus*. The nest of this bird is notable for the extreme neatness and compactness of its structure, for it can endure a vast amount of careless handling, and still retain its beautiful contour. A specimen in my collection was taken from the banks of a river near Natal, and was suspended from two reeds, so as to hang over the water, and at no great distance from the surface.

The whole structure is apparently composed of the same plant, namely, a kind of small reed, but the materials are taken from a different portion of the plant, according to the part of the nest for which they are required. The whole exterior, as well as the walls. are made of the reed-stems, woven very closely together, and being of no trifling thickness. There is a considerable amount of elasticity in the structure, and the whole nest is so strong that it might be kicked down stairs, or be thrown from the top of the Monument, without much apparent deterioration. The interior, however, is constructed after a very different fashion. Instead of the rough, strong workmanship of the exterior, with its reed-stems interlacing among each other, as if woven by human art, and its pale yellow hue, the inside exhibits a lining of flat leaves, laid artistically over each other so as to form a soft, smooth resting-place, but not interlacing at all, being held in their place by their own elasticity. Their colour is of a pale bluish grey, and the contrast which they present to

THE TAILOR BIRD.

the exterior is very strongly marked. In size the nest is about as large as an ordinary cocoa-nut—not quite so long, though broader.

The wonderful little bird, whose portrait is accurately given in the accompanying illustration, is popularly known by the appropriate title of TAILOR BIRD, its scientific name being *Or-*

thotomus longicaudus. The manner in which it constructs its pensile nest is very singular. Choosing a convenient leaf, generally one which hangs from the end of a slender twig, it pierces a row of holes along each edge, using its beak in the same manner that a shoemaker uses his awl, the two instruments being very similar to each other in shape, though not in material.

When the holes are completed, the bird next procures its thread, which is a long fibre of some plant, generally much longer than is needed for the task which it performs. Having found its thread, the feathered tailor begins to pass it through the holes, drawing the sides of the leaf towards each other, so as to form a kind of hollow cone, the point downwards. Generally a single leaf is used for this purpose, but whenever the bird cannot find one that is sufficiently large, it sews two together, or even fetches another leaf and fastens it with the fibre. Within the hollow thus formed the bird next deposits a quantity of soft white down, like short cotton wool, and thus constructs a warm, light, and elegant nest, which is scarcely visible among the leafage of the tree, and which is safe from almost every foe except man.

The Tailor Bird is a native of India, and is tolerably familiar, haunting the habitations of man, and being often seen in the gardens and compounds, feeding away in conscious security. It seems to care little about lofty situations, and mostly prefers the ground, or lower branches of the trees, and flies to and fro with a peculiar undulating flight. Many species of the same genus are known to ornithologists.

THE tailor bird is not the only member of the feathered tribe which sews leaves together in order to form a locality for its nest. A rather pretty bird, the FAN-TAILED WARBLER (*Salicaria cisticola*) has a similar method of action, though the nest cannot be ranked among the pensiles.

This bird builds among reeds, sewing together a number of their flat blades in order to make a hollow wherein its nest may be hidden; but the method which it employs is not precisely the same as that which is used by the tailor bird. Instead of passing its thread continuously through the holes, and thus sewing the leaves together, it has a great number of threads, and makes a knot at the end of each, in order to prevent it from being pulled through the hole. A description and beautiful figure of this bird may be seen in Gould's 'Birds of Europe,' vol. ii.

CHAPTER XII.

PENSILE BIRDS (*continued*).

Australian Pensiles — The YELLOW-THROATED SERICORNIS—Its habits — Singular position for its nest—Conscious security—The ROCK WARBLER—Shape and locality of its nest—The SINGING HONEY-EATER and its nest—The myall or weeping acacia—Various materials—The PAINTED HONEY-EATER, its habits and nest—The WHITE-THROATED HONEY-EATER and its habits—Its curious nest—Locality of the nest—The SWALLOW DICÆUM—Its song and beauty of its plumage—The nest, its materials, form, and position—The HAMMOCK BIRD—Singular method of suspending the nest.

SOME very remarkable instances of pensile birds' nests are found in Australia, and for many of them we are indebted to the careful and painful research of Mr. J. Gould, from whose skilful works on ornithology several illustrations have been, by permission, copied.

A very curious instance is found in the nest of the Yellow-throated Sericornis (*Sericornis citreogularis*), a rather pretty, but not a striking bird. The general colour is simple brown, and, as its name imparts, the throat is of a citron-yellow. The only remarkable point in the colour, beside the yellow throat, is a rather large patch of black, which envelopes the eye and passes down each side of the neck, nearly as far as the shoulders. It is the largest of its genus, and, although not rare, is seldom seen except by those who know where to look for it, as it is scarcely ever observed on the wing, but remains among the thick underwood, flitting occasionally between the branches, but mostly remaining on the ground, where it pecks about in search of the insects on which it feeds.

The reason for its mention in this work is the singular structure of its nest, which is described by Mr. Gould in the following words :—

'One of the most interesting points connected with the history of this species is the situation chosen for its nest.

'All those who have travelled in the Australian forests must have observed that, in their more dense and humid parts, an

PTILOTUS SONORUS. ENTOMOPHILA PICTA. ENTOMOPHILA ALBOGULARIS.
SERICORNIS CITREOGULARIS. ORIGMA RUBRICATA.

atmosphere peculiarly adapted for the rapid and abundant growth of mosses of various kinds is generated, and that these mosses not only grow upon the trunks of decayed trees, but are often accumulated in large masses at the extremities of the drooping branches. These masses often become of sufficient size to admit of the bird constructing a nest in the centre of them, with so much art that it is impossible to distinguish it from any of the other pendulous masses in the vicinity. These bunches are frequently a yard in length, and in some places hang so near the ground as to strike the head of the explorer during his rambles; in others, they are placed high up on the trees, but only in such parts of the forest where there is an open space entirely shaded by overhanging foliage. As will be readily conceived, in whatever situations they are met with, they at all times form a remarkable and conspicuous feature in the landscape.

'Although the nest is constantly disturbed by the wind, and liable to be shaken when the tree is disturbed, so secure does the inmate consider itself from danger or intrusion of any kind, that I have frequently captured the female while sitting on her eggs, a feat that may always be accomplished by carefully placing the hand over the entrance—that is, if it can be detected, to effect which, no slight degree of close prying and examination is necessary.

'The nest is formed of the inner bark of trees, intermingled with green moss, which soon vegetates; sometimes dried grasses and fibrous roots form part of the materials of which it is composed, and it is warmly lined with feathers. The eggs, which are three in number, and much elongated in form, vary considerably in colour, the most constant tint being a clove-brown, freckled over the end with dark umber-brown, frequently assuming the form of a complete band or zone; their medium length is one inch, and their breadth eight lines.'

If the reader will bear in mind the remarkable shade of this and a few other nests, he will see, in a future page, how wonderful is the resemblance between the pensile nests of birds and insects.

PENSILE birds do not always suspend their nests to the branches of trees, but in some instances choose exactly the localities which appear to be the most unsuited for the purpose. Still keeping to Australia, we may find a most wonderful example of a pensile nest near mountain courses. The bird which makes it is called, indifferently, the ROCK WARBLER, or the CATARACT BIRD (*Origma rubricata*), because it is always found where water-courses rush through rocky ground. So attached is the bird to these localities, that it is never seen in the forest, nor ever has been observed to perch upon a branch. The generic name, *Origma*, is derived from a Greek word, signifying a rock or a precipice, and is more appropriate than are many scientific titles.

It is a small bird, no larger than our sparrow, and is soberly coloured, the general hue being brown, relieved by a dull red on the breast, something like that of the female robin. It has a melodious though not very powerful note; but its chief claims to admiration are founded upon the extraordinary nest which it builds. In general shape this nest somewhat resembles a claret jug without a handle, having a long, slender neck and a globular and suddenly-rounded bulb.

It is suspended from the rocks in sheltered places, and whenever an overhanging ledge of rock affords protection from the elements, there the strange nests may be found. Just as the martins take a fancy to some favourite spot, and build whole rows of nests on one side of some particular house, utterly disdaining neighbouring houses, which, to all appearance, afford exactly the same advantages, so do the Rock Warblers affect some particular rock, and hang their nests by dozens in close proximity to each other. The material of the nest is the long moss which is plentiful in the country; and, as it may be seen from the illustration, the entrance is near the centre of the rounded bulb. In consequence of the material of which the nest is constructed, it is very rough on the exterior, though smooth and comfortable within.

A MOST beautiful pensile nest is made by the SINGING HONEY-EATER (*Ptilotus sonorus*), a species which is spread over a large portion of Australasia.

Here we have another example of an Australian singing bird, for the melody of this creature is so loud, so full, and so rich in tone, that Mr. Gould compares it to that of the missel thrush. It is a soberly-coloured bird, though easily identified, the back being pale brown, the top of the head yellow, and a deep black patch passing over the eye and turning downwards along the side of the neck. It is a lively bird, as are all those feathered creatures which feed chiefly on insects, and even in mid-winter its melodious song may be heard in full vigour.

There is a very common tree in Australia, popularly called the myall, known to scientific botanists as *Acacia pendula*. The twigs of the tree are long and very slender, and the leaves are so narrow and delicate that at a little distance they look more like grass-blades than the leaf of a tree. The reader may remember that this is a characteristic of all drooping or 'weeping' trees, the leaf and the twig being slender in proportion to each other. The weeping birch and the weeping willow of our own country are good examples of this peculiarity.

Thus, as both the leaves and the twigs of the myall are extraordinarily long and slender, the tree is chosen by many birds which build pensile nests, as will be seen in the course of this volume. It seems a tree that was made for the express purpose, because the long and slender twigs serve the double purpose of affording a firm attachment for the nest and suspending it where no ordinary foe can reach it, while the delicate leaves give their aid in fastening the nest to the twigs, and at the same time serve to conceal the structure from prying eyes.

Although the general structure of the nest is the same in all parts of the country, the materials necessarily differ. In New South Wales, the external shell of the nest is formed of very fine dry stalks, not thicker than twine, while the lining is composed of fibrous roots, matted together with spiders' webs. It is fastened by the rim to the twigs, and as a few of the slender twigs occasionally are interwoven into the nest, it hangs quite securely. In Western Australia, the nest is made of grasses, which, although green when first woven, become white and dry in a short time. The grass is mingled with the hair of the Kangaroo and the fur of some phalangist, vulgarly called opossum, which serve to mat the grass together, and to make it impervious to the wind and rain; and the interior is neatly lined with grasses and vegetable down.

THERE is another of these pretty birds, called the PAINTED HONEY-EATER, on account of the variety of its colouring. Its scientific name is *Entomophila picta.* The general colour of this handsome bird is rich brown above, with the exception of a yellow patch on the base of the tail, and white, slightly spotted, below. A characteristic mark of the species is a little patch of pure white just by the ears.

This handsome species inhabits the interior of New South Wales, and does not confine itself merely to a diet of sweet juices, but feeds much on small insects. The generic title, Entomophila, is composed of two Greek words, which signify insect-lover, and is given to this bird, and several other Honey-Eaters, on account of their insect-eating habits. The birds are extremely active, and devote much of their time to the pursuit of insects on the wing, in which occupation they have a great resemblance to our well-known fly-catcher. They sit on a branch, keeping a careful watch, and whenever an insect passes near, they dart into the air, catch it, and return to their post. They are generally seen in pairs, and are very playful, chasing each other merrily, and spreading their tails so as to show the white colour. When on the wing, they are so like the common goldfinch that they might easily be mistaken for that bird, the patchy distribution of the colour, and the white spot on the face, adding greatly to the resemblance.

The material of which the nest of the Painted Honey-Eater is composed is fine fibrous roots, interwoven very artfully, but loosely, and being of so frail a structure, that much care is required to remove it without damage. It is fastened by the rim to the delicate twigs of the beautiful weeping acacia (*Acacia pendula*), whose long lanceolate leaves droop over and nearly cover it. It is a very small nest in proportion to the size of the bird.

STILL keeping to the same interesting family of birds, we find among the pensile builders another species of Honey-Eater.

The WHITE-THROATED HONEY-EATER (*Entomophila albogularis*) is rather like the Painted Honey-Eater, being brown above, white below, and having a yellow patch on the base of the tail. It is, however, easily distinguished from its congener by the peculiarity from which it derives its name—viz. a large patch of pure white in the front of the throat, extending as far as the eyes. The top of the head is greyish blue, and the breast is buff.

It is a lively, active little creature, ever on the move, and delighting to flit from branch to branch, but not caring to make long flights. As it flies from one bough to another, it utters a musical little song, much like that of the goldfinch, and continues to sing for a considerable time. It detests wind, and is mostly seen in the thick bush, and loves to frequent the masses of mangroves which edge bays and creeks, because the air is comparatively still. In these places may be found its curious nest, which is about as large as a breakfast-cup, and very much of the same shape. It is made of the delicate paper-like bark of the Melaleucæ, and various vegetable fibres, with which it is ingeniously hung to the branches. The broad, thin bark causes it to be very smooth on the exterior. For the lining, the bird is not indebted to any animal or bird, but uses grass-blades, which are neatly laid, and form a soft resting-place for the eggs.

The nest is placed very low, being often found scarcely two feet from the water, in that point resembling the nest of the African weaver birds, which have already been described. It is always hung near the extremity of a branch, and invariably is so placed as to be under the protection of a spray of leaves, which act as a roof whereby the rain is thrown off.

THERE is a genus of very small birds, called Dicæum, which is spread over many parts of the world, and finds several representatives in Australia. All are interesting birds; but as the present work only treats of birds as the architects of their nests, it is necessary to select one which builds a pensile habitation. This is the SWALLOW DICÆUM (*Dicæum hirundinaceum*), a bird scarcely as large as our common wren, and glowing with brilliant colours, the whole of the upper part being deep, glossy blue-black; the throat, breast, and under tail-coverts of a fiery scarlet; and the abdomen pure white. It has a very sweet

SWALLOW DICÆUM.

though low and inward note, so faint as scarcely to be audible from the tops of the trees, but continued for a long time together.

Artificial aids to vision are required in order to watch the habits of the Dicæum, for it loves the tops of the tallest trees, where its minute body can scarcely be seen without the assistance of glasses. The Casuarinæ are favourite trees with this bird, which is fond of flitting about the branches of a parasitic plant called loranthus, which bears viscid berries. It is not precisely known whether the bird haunts the loranthus for the sake of the berries or of the insects, but as the Dicæum is one of the insect-eaters, the latter supposition is probably correct.

It is very seldom if ever seen on the ground, and its flight among the upper branches is quick, sharp, and darting.

The nest of the Swallow Dicæum is as pretty as its architect, and its ordinary shape can be seen in the accompanying illustration, though the plain black and white of a wood engraving can give but little idea of its full beauty. In colour it is nearly pure white, being made of the cotton-like down which accompanies and defends the seeds of many plants, and this material is so artfully woven that the nest almost looks as if it were made from a piece of very white cloth. It is always purse-like in form, though its shape is slightly variable, and is suspended by the upper portion to the twigs at the very summit of the tree. Generally it hangs its nest upon the parasitic plant which has already been mentioned, but it often selects the Casuarinæ, or the delicate twigs of the myall or weeping acacia, for that purpose. The average number of eggs is five, and their colour is greyish white thickly powdered with small brown specks. Their length is about three-quarters of an inch, and their breadth rather less than half an inch.

ON the next page is a portrait of one of the Honey-Eaters, called the LANCEOLATE HONEY-EATER (*Plectorhynchus lanceolatus*), on account of the shape of its feathers. It is not a brilliantly coloured bird, its hues being only brown and white, diversified by a black line down the middle of each feather.

The wonderful nest of this bird was found by Mr. Gould on the Liverpool Plains, overhanging a stream, and being a beautiful example of the pensiles. The materials of which it is made are grass and wool, intermingled with the pure white cotton of certain flowers. As the reader may see, by reference to the illustration, it is hung from a very slender twig, and only suspended at opposite extremities of the rim, the tree selected

LANCEOLATE HONEY-EATER. (*Plectorhynchus lanceolatus.*)

being the myall, or weeping acacia. The nest is rather small in proportion to the bird, and is very deep, so that when the mother is sitting on her eggs, or brooding over her young, she is obliged to pack herself away very carefully, her tail projecting at one side of the nest and her head at the other.

CHAPTER XIII.

PENSILE BIRDS (*continued*).

American Pensile Birds—Humming Birds, and the general structure of their nests—The LITTLE HERMIT, its colour, habits, and nest—The GREY-THROATED HERMIT and its hardihood—The PIGMY HERMIT and its seed-nest—The WHITE-SIDED HILL STAR—Curious method of suspending its nest—The SAWBILL and its singular nest—Habits of the Sawbill—The BRAZILIAN WOOD NYMPH — Use made of its plumage and nest — The BALTIMORE ORIOLE—Reason for its name—Its beautiful nest, and curious choice of materials — The ORCHARD ORIOLE, or BOB-O'-LINK — Various forms of nest — Why called Orchard Oriole — The CRESTED CASSIQUE, its size, form, and colours—Its remarkable nest—The GREAT CRESTED FLY-CATCHER, and its use of serpent-sloughs—The RED-EYED FLY-CATCHER, or WHIP-TOM-KELLY—Low elevation of its nest—The WHITE-EYED FLY-CATCHER, its nest, and fondness for the prickly vine—The Asiatic Pensiles —The BAYA SPARROW — Its colour and social habits — Singular form of the nest.

HAVING now taken a cursory glance at the pensile nests constructed by the feathered inhabitants of Africa and Australia, we again cross the sea and come to America. There are many pensile builders among American birds, and the chief among them are the exquisite little creatures called the HUMMING BIRDS, which are peculiar to America and her islands.

Among the multitudinous species of this wonderful group of birds are very many examples of pensile nests, that mode of structure being, indeed, the rule, and any other the exception. As is the case with the nests of the Australian birds, some are suspended from twigs, others from rocks, and others again from leaves, the last-mentioned plan being the most common.

Our first example of the pensile Humming Birds is the beautiful species called the LITTLE HERMIT (*Phaethornis eremita*).

The nest which is here figured was attached to the very extremity of the leaf, so that the long tail hung down freely.

The materials of which it was composed were the silky fibres of plants, the cotton-like down of seed vessels, and some other substance, which is supposed to be fungus, and is of a woolly texture. All these materials were interwoven with spiders' web,

LITTLE HERMIT. (*Phaëthornis eremita.*)

by means of which the nest was attached to the leaf at the end of which it swings. The bird almost invariably chooses some dicotyledonous leaf for its pendant home.

Other nests made by birds of the same genus are worthy of a passing mention.

First, there is the pretty nest of the GREY-THROATED HERMIT (*Phaethornis griseogularis*), a very tiny bird, of comparatively sober plumage, reddish brown being the predominant hue. This species is found in Ecuador, and is seen at an elevation of six thousand feet above the level of the sea. Indeed, the depth of cold which these fragile little beings can endure is really surprising, many species being found only on the highest mountains, and one bird, the Chimborazian Hill Star, inhabiting a zone that is never less than twelve thousand feet, and seldom more than sixteen thousand, above the level of the sea. Immediately above the last-mentioned elevation the line of perpetual snow begins, and though the bird can exist just below it, the absence of vegetation prevents it overpassing that line.

The nest of the Grey-throated Hermit is made of moss fibres and the same silken threads that have already been mentioned, and is fastened to a leaf. It does not, however, hang from the extremity, but is fastened against the side of the leaf, and its tail, if we may so call the lengthened appendage, is not free, but attached to the leaf in the same manner as the nest.

Another species, *Phaëthornis Eurynome*, makes its nest of the tendrils of certain creepers, together with delicate root-fibres, and attaches it to the leaf of some palm by means of cobwebs.

Our last example of this group is the tiny species called the PIGMY HERMIT (*Phaethornis pygmæus*), a pretty little creature, though scarcely a brilliant one, and decorated with green-bronze above and warm red below. The nest of this species is fastened to a leaf, like that of the grey-throated hermit, and is also deep and cup-shaped, with an appendage so long as to give the whole nest a shape resembling that of a funnel. It is remarkable for the great use of which this little architect makes of seeds, the exterior being covered with downy seeds, and the interior lined with similar down, and the delicate fibres of flowering plants.

IN the accompanying illustration may be seen figures of the nests made by three different species of humming birds, each of which is remarkable for some peculiarity of structure, though they are all pensile.

The first of these nests is that which is made by the WHITE-SIDED HILL STAR (*Oreotrochilus leucopleurus*); a native of the Andes of Acoucagua, inhabiting a zone of very great elevation, seldom being seen less than ten thousand feet above the level of the sea. With the exception of a bright emerald-green gorget,

SAWBILL HUMMING BIRD. BRAZILIAN WOOD NYMPH. WHITE-SIDED HILL STAR.

it is rather a dull-coloured bird, the prevailing hue being brown. The nest is shaped something like a hammock, not unlike that of the lanceolated honey-eater, described and figured on page 136, and is fastened, not to a twig or a leaf or a branch, but to the side of a rock, being suspended by one side, so as to leave the remainder free.

As is the case with the generality of humming birds' nests, cobwebs are employed for the purpose of fastening the structure to the object to which it hangs. The materials of which the nest is made, are chiefly moss, down, and feathers, the feathers being profusely stuck on the outside.

THERE is a very remarkable nest made by one of these birds, called the SAWBILL HUMMING BIRD (*Grypus nævius*), because the slender bill is notched in a saw-like fashion on the edges of both mandibles. These serrations do not reach along the whole bill but only to a short distance from the tip.

The nest of the Sawbill is made of fine vegetable fibres, woven together so as to look like an open network purse, the outer walls being so loosely made as to permit the eggs and lining to be visible. Leaves, mosses and lichens are also woven into the nest, and are packed rather tightly under the eggs. The edge, however, is always left loose. The nest is suspended at the end of some leaf, usually that of the palm.

Mr. Gould mentions that the bird is found in the depths of virgin forests, and is most plentiful about thirty miles from Nova Fribergo, in the months of July, August, September, and part of October. It is generally seen darting round the orchidaceous plants which flower so richly in that fertile climate, and is a rather noisy bird, uttering loud and piercing cries, and making a great whirring sound with its wings as it dashes through the air. It is very strong and energetic on the wing, and is seldom seen to alight. That the Sawbill feeds on insects has been satisfactorily proved, by the presence of small beetles in the throat of newly killed birds; and to judge by its actions, the hovering flight and frequent stoop like that of the falcon, the bird feeds also on flies and other winged insects.

ALTHOUGH it is necessarily impossible to describe or even enumerate one tithe of the interesting nests made by humming birds, I must cursorily mention one or two more of the most curious examples. One of these birds is the BRAZILIAN WOOD

NYMPH (*Thalurania glaucopis*), a species which is perhaps more persecuted than any other, its singular beauty causing its plumage to be sought after

The feathers on the crown of the head and front of the throat are of the most lovely azure, and are largely used by the inmates of several convents at Rio Janeiro for the purpose of being made into the beautiful feather flowers which the nuns manufacture so skilfully. Thousands of these birds are slaughtered merely for the crest and gorget, but so prolific are they. and so ingeniously do they hide their nests, that the persecution of many years has scarcely diminished their numbers. Moreover, fortunately for the preservation of the species, the colours of the female are so dull and sober, that her feathers are of no value, and she is allowed to escape the fate that befalls the more brightly coloured male. It is a lively little bird, and when alarmed utters a hurried cry, sounding like the word, 'Pip, pip, pip,' very sharply pronounced.

The nest of the Brazilian Wood Nymph is exceedingly pretty, and is hung to the tip of some delicate twig, generally that of one of the creeping plants which trail their long stems so luxuriantly over the branches of the great forest trees. The walls of the nest are made of vegetable fibres, generally taken from the fruit of some palm, and upon the outside are fastened many patches of flat lichen, so that the whole nest, which is very long in proportion to its width, may easily escape detection.

Two differently-shaped specimens are given in the accompanying illustration, in order that they may be compared with each other.

The first in order is that of the BALTIMORE ORIOLE (*Yphantes Baltimore*), a pretty bird, coloured with orange and black in bold contrast to each other. Its name is derived, not from any particular locality, but from the orange and black of its plumage, those being the heraldic colours of Lord Baltimore, formerly proprietor of Baltimore. It does not receive the full colouring until its third year, the orange hues being simply yellow at the end of the second year, and having no red in them until the last moult is completed. So far, indeed, is it from belonging to any particular locality, that it is spread over a very wide range of country, inhabiting the whole of America

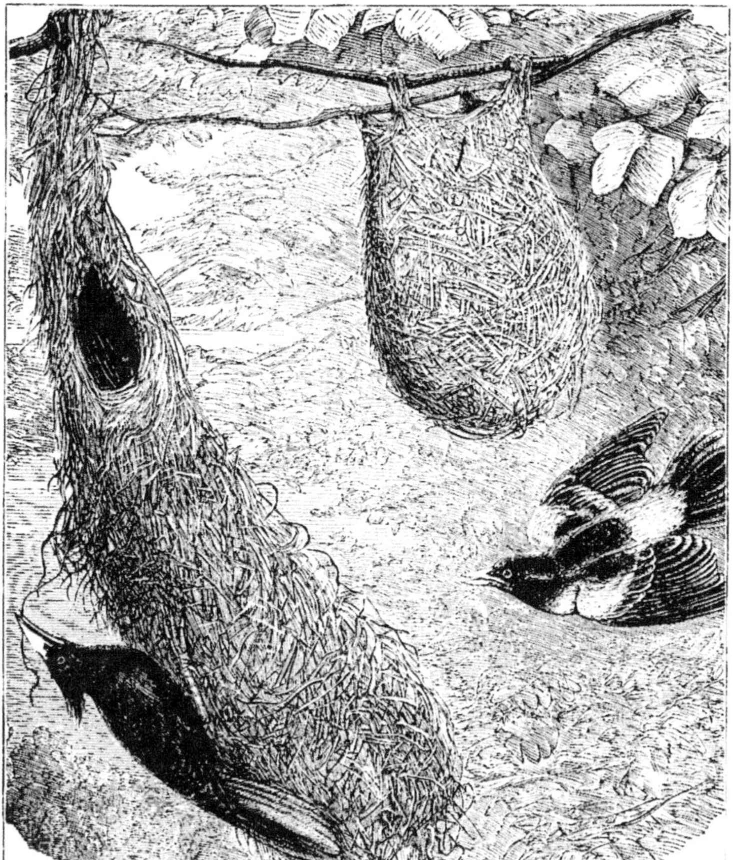

CRESTED CASSIQUE. BALTIMORE ORIOLE.

from Canada to Brazil. The Baltimore Oriole goes by many names; some, such as Golden Robin and Fire Bird, being in allusion to its plumage, and others, such as Hang-nest and Hanging Bird, from the beautiful pensile nest which it makes.

The general shape of these nests is much the same in every specimen, and a good idea of it may be formed from the illustration, which was taken from a nest in my own possession. It is almost entirely made of vegetable fibres, and is so strongly constructed, that, although it had been knocked about for some years in the neglected spot whence I rescued it, and was once crushed into a shapeless mass at the bottom of a wine hamper by a careless servant, and covered with soot and dust, it has retained its form, and shows perfectly well how the fastening to the branches was managed.

The materials of the nest are, however, extremely variable, the bird having a natural genius for nidification, and being always ready to take advantage of any new discovery in architecture. One of these nests, described by Wilson, was deeper in proportion than the specimen which has been figured, being five inches in its widest diameter and seven in depth, the opening being contracted to two and a half inches. Various materials, such as flax, tow, hair, and wool, were woven into the walls, which were strengthened by horsehairs, some two feet in length, sewn through and through the fabric. Cow's hair was also employed for the bottom of the nest, and, like the walls, was sewn together with long horsehairs.

The same writer remarks, that ' so solicitous is the Baltimore to procure proper materials for his nest, that in the season of building, the women in the country are under the necessity of narrowly watching their threads that may chance to be out bleaching, and the farmer to secure his young grafts; as the Baltimore, finding the former, and the strings which tie the latter, so well adapted for his purpose, frequently carries off both. Or, should the one be over heavy, and the other too firmly tied, he will try at them for a considerable time before he gives up the attempt. Skeins of silk and hanks of thread have often been found, after the leaves were fallen, hanging round the Baltimore's nest, but so woven up and entangled as to be entirely irreclaimable.

A closely allied species, the Orchard Oriole, or Bob-o'-link (*Xanthornis varius*), is equally notable for its skill in nest-building—if such a word may be used of a structure which is begun at the top and carried downwards, after the fashion employed in Laputa.

It is a pretty bird, but not so pretty as the Baltimore Oriole, and the tints are very differently disposed, scarcely any two individuals having the colours in exactly the same places. Like the Baltimore Oriole, it is extremely variable in different stages of its existence, the young male bearing great resemblance to the mature female, and not attaining its full beauty until its third year. When adult, the whole of the head, neck, upper part of the back, breast, wings, and tail, are deep black, and a rich ruddy chestnut hue occupies the remainder of the breast, the under parts of the body, and part of the wing-coverts, some of which are tipped with white. The young male and the adult female are yellowish olive above, instead of black, with brown wings, and yellow on the breast and abdomen; while the male of the second year has much the same colours, but is known by a patch of black over the head and on the throat, together with a few chestnut feathers on the flanks and abdomen. It is smaller than the Baltimore Oriole, and more slenderly made.

The nest of this bird is almost as variable in structure as is its architect in colour, its form being accommodated to the situation in which it is placed. When fastened to a tolerably stout branch, its depth is less than its diameter, and it is firmly tied in several directions to prevent the wind from upsetting it. But when it is slung to a long and slender branch, over which the wind has great power, and which is swung to a distance of fourteen or fifteen feet in a smart breeze, the nest is made of much greater depth, and is of a lighter construction. The weeping willow is a favourite tree with this bird, as the drooping leaves conceal the nest effectually, and the delicate twigs can be gathered together so as to support the entire circumference of the entrance.

On the left hand of the Baltimore Oriole's nest is represented a very curious structure swaying in the wind, long, purse-like, and having the entrance near the top. This is the nest of the Crested Cassique, or Crested Oriole (*Cacicus cristatus*), and the bird itself is seen clinging to the lower part of the nest.

There are several species of Cassiques, all of which are natives of tropical America, and build nests of a similar structure. The Crested Cassique is the largest of the genus, equalling the common jackdaw in size, and its nest is larger and more striking than that of any other species. It loves the tallest trees, and may be seen actively traversing the branches in search of food, pecking here and there in haste as it trips along, or passing from one tree to another with a rapid darting flight, snapping at insects as it dashes through the air. Like the preceding species, it is fond of human society, and builds its pensile nest close to the habitation of man, so that its customs can be easily watched.

The bird is a handsome creature, the greater part of the body being rich chocolate, the wings dark green, and the outer tail-feathers bright yellow, this colour being displayed conspicuously as the bird flies, particularly when it makes a sharp turn in the air and is obliged to spread its tail-feathers rapidly. The beak of this species is very remarkable, being of a green colour, and extending far up the forehead. The head is adorned with a long pointed crest, from which its popular name of Crested Oriole is derived. In some favoured spots these birds are quite plentiful, producing a beautiful effect, as the variegated plumage gleams among the foliage, while the bird is engaged in its active quest after food.

The nest of the Crested Cassique is of great length, and, as may be seen by the illustration, has the entrance like that of a pocket. The opening is rather small when compared with the size of the nest itself, and the bird always dives head foremost into its home, its yellow tail flashing a last golden gleam before it disappears. The nest is strongly built, and the materials are rather coarse, not in the least resembling the delicate and neatly rounded fibres of which many of the weaver nests are made. These nests often exceed a yard in length. and owing to their great size, are very conspicuous, as the wind sways them backwards and forwards from the bough.

Before leaving the American pensile birds, we must briefly notice one or two other species. The Flycatchers of all countries are generally notable for the beauty or eccentricity of their nests, one of the oddest being that of the Great Crested Flycatcher of America, which always uses the cast slough of snakes when building its nest. The reason no one seems to know, though several opinions have been offered; one person thinking the snake-slough is peculiarly grateful to the young birds which are intended to lie upon it; and another, that the presence of the cast slough acts as a scarecrow, and frightens away obnoxious birds. One conjecture is as good as another, and both are absurdly bad.

The species which we have now to notice is the Red-Eyed Flycatcher (*Muscicapa olivacea*) popularly known as 'Whip-Tom-Kelly,' from its peculiar articulate cry, which is said to bear a strangely exact resemblance to the words 'Tom Kelly, Whip-tom-kel-ly,' and is uttered so loudly and briskly, that it can be heard at a considerable distance. It inhabits a tolerably wide range of country, being found from Georgia to the St. Lawrence, and in many parts is plentiful.

The nest of the Red-Eyed Flycatcher is small and very neatly made, and, contrary to the usual custom of pensile nests, is placed near the ground, seldom at a height of more than five feet. Bushes and dwarf trees, such as dogwood or saplings, are usually chosen by the bird when it looks about for a branch wherefrom to hang its nest. A wonderful array of materials is employed by the feathered architect, which makes use of bits of hornets' nests, dried leaves, flax-fibres, strips of vine bark, fragments of paper and hair, and binds all these articles firmly together with the silk produced by some caterpillars. The lining is made of fine grasses, hair, and the delicate bark of the vine.

The nest is wonderfully strong, so compact indeed, that after it has served the purpose of its architect, it is usurped by other birds in the following year, and saves them the trouble of building entire nests of their own. Even the mammalia receive some benefit from the nest, for the field-mouse often takes possession of it, and rears its young in the pensile cradle.

An allied species, the WHITE-EYED FLYCATCHER (*Muscicapa cantrix*), builds a very pretty pensile nest, and uses so much old newspaper in the construction of its home, that it has gone by the name of the POLITICIAN. The other materials used in the structure of the nest are bits of old rotten wood, vegetable fibres, and other light substances, woven together with wild silk, and the lining is mostly of dried grasses and hair.

The form of the nest is nearly that of an inverted cone, and it is suspended by part of the rim to the bend of a species of smilax, that is popularly called the prickly vine, and which grows in low thickets. The bird is very fond of this smilax and rarely chooses any other tree for the reception of its nest, so that the home of the White-Eyed Flycatcher is not very difficult to find; moreover, the bird is so jealous and so bold when engaged in rearing its young, that it betrays the position of the nest by scolding angrily as soon as a human being approaches the thicket, and by dashing violently at the intruder with impotent rage.

As we are near the end of our list of pensile birds, we must turn to Asia for a specimen as remarkable as any which has yet been mentioned. This is the nest of the BAYA SPARROW, sometimes called the TODDY BIRD, a native of several parts of India, and found in Ceylon.

As may be seen by the frontispiece, the nests are variable in shape, and hang close to each other; indeed, the birds are very sociable in all their manners, and fly about in great numbers, flocks of thousands flitting among the branches and displaying their pretty plumage to the sun. They have no song, and can only chirp in a monotonous manner; but the want of song finds its compensation in the brilliancy of the plumage, which is mostly bright yellow, the wings, back, and tail being brown. They are particularly fond of the acacias and date-trees, and choose the branches of those trees for the suspension of their nests.

Sometimes the nest is only made for incubation, sometimes it is intended merely as an arbour in which the male sits while the female incubates her eggs, and sometimes it consists of the nest and arbour united, producing a most curious effect. This 'arbour,' in fact, serves precisely the same purpose as the supplementary nest of the pinc-pinc and other birds which have already been described.

The frontispiece represents a group of Baya Sparrows' nests, taken from a photograph. The photograph was sent to the Zoological Society by C. Horne, Esq., who furnished the following valuable account of the mode of nest-building; it appeared, together with a lithograph of the tree and nests, in the 'Proceedings of the Zoological Society, 1869.'

'This morning (July 7, 1865) as I passed our solitary palm tree (*Phœnix dactylifera*) in the field, I heard a strange twittering overhead, and looking up, saw such a pretty sight as I shall never forget.

'In this tree hung some thirty or forty of the elegantly formed nests of woven grass of the Baya bird, so well known to all. The heavy storms of May and June had torn away many, and damaged others so as to render them, as one would think, past repair. Not so thought the birds, for a party of about sixty had come to set them all in order.

'These little birds are about the size of a sparrow, and have yellow in their crests, and are darker about the wings, being paler below, with shortish tails. The scene in the tree almost baffles description. Each bird and his mate thought only of their own nest. How they selected it I know not, and I should much like to have seen them arrive. I suppose the sharpest took the best nests, for they varied much in condition. Of some of the nests two-thirds remained, whilst others were nearly blown away. Some of the birds attempted to steal grass from other nests, but generally got pecked away.

'As the wind was blowing freshly, the nests swung about a good deal, and it was pretty to see a little bird fly up in a great hurry with a long bit of grass in his beak. He would sit outside the nest, holding on by his claws with the grass under them. He would then put the right end into the nest with his beak, and the female inside would pull it through and put it out for him again, and thus the plaiting of the nest went on. All this was done amidst great chattering, and the birds seemed to think it great fun. When a piece was used up one would give the other a peck, and he or she would fly off for more material, the other sitting quietly till the worker returned. Nests in every stage of building afforded every position for the bird, who seemed at home in all of them. The joy, the life, the activity, and general gaiety of the birds I shall never forget.

'August 18.—Noticed to-day how the birds obtained their grass. The little bird alights at the edge of the high, strong Seenta grass (*Andropogon euripeta?*) with its head down, and bites through the edge to the exact thickness which it requires. It then goes higher up on the same blade of grass, and having considered the length needed, bites through it again. It then seizes it firmly at the first notch and flies away. Of course the strip of grass tears off, and stops at the notch. It then flies away with the grass streaming behind it. As the edge of the grass is much serrated, the bird has to consider and pass it through the work the right way.

'In some instances the male continues to build for amusement after the nest is finished, not only elongating the tubular entrance, but also making a kind of false nest.'

CHAPTER XIV.

PENSILE INSECTS.

The Hymenoptera—The TATUA, or DUTCHMAN'S PIPE—Structure and Shape of its Nest—Firmness of the Walls—Average number of Cells in each Tier—The NORWEGIAN WASP—Structure and Locality of its Nest—The CAMPANULAR WASP and the NORTHERN WASP—Honey Wasps, the general characteristics of their Nests—The MYRAPETRA—Its singular Nest—Structure of the Walls and use of the Projections—The NECTARINIA—Why so called—Locality of the Nest—Size of the Insect—Ichneumon Flies—Different species of MICROGASTER, and their Habitations—The ATLAS MOTH—The HOUSEBUILDER MOTH and its movable Dwelling—The TIGER MOTH and its Hammock—The BARNET MOTH and its Cocoon—The OAK EGGAR and LITTLE EGGAR MOTHS—Various Leaf-rollers—Suspended Cocoon—LEAF-BURROWERS and their Homes—The SPIDER.

WE now leave the birds, and proceed to the insects which make pensile nests. Some of them, such as those which will be first described, do not become pensile architects until they have attained their perfect state; while many others form their nests, either as a place of refuge during their larval life, or as an asylum in which they can rest while in the transition state of pupa.

Just as the Hymenoptera are the best burrowers, so are they the best insect artizans when the nests are suspended, and we shall therefore take them first in order.

IN the accompanying illustration may be seen two specimens of a remarkable pensile nest that is made by a wasp called *Tatua morio*, an insect which is notable for having the basal segment of the abdomen narrowed into long and slender foot-stalks, not unlike that of the Eumenes, and others.

The nest of this species is made of the papery substance used by many wasps, except that the material is so hard and smooth as to resemble white cardboard. The general form of the nest is shown in the engraving, being somewhat like a sugarloaf, *i.e.* a round-topped cone with a flat bottom. It is found in several parts of Central America; and in Guiana the nest goes by the popular name of 'the Dutchman's pipe,' being supposed to bear, in shape and dimensions, some resemblance

TATUA MORIO.

to the pipe-bowl celebrated by Washington Irving. The exterior walls are so hard, firm, and smooth, that they can withstand any vicissitudes of weather, neither the fierce storms that blow in those regions, nor the torrents of rain which occasionally fall, having any power over an edifice so well protected.

The tiers of cells are variable in number; a rather remarkable fact, as the floors are made before the cells are built. In a good specimen of this nest in the British Museum there are only four tiers of cells. How many tiers are completed before the insects begin to affix cells to them, or whether the cells are made as soon as the floors are finished, are two points in the history of this wasp which have not yet been decided. These floors extend completely to the walls, to which they are fastened on all sides, and the insects gain admission to the different floors by means of a central opening which runs through them all.

In Mr. Waterton's museum, at Walton Hall, are several specimens of these nests, one of which is opened so as to show the interior, as well as the central aperture, the whole of the bottom being cut away and raised like the lid of a box. The substance of this nest resembles thin brownish pasteboard, and, as is the custom with most of the wasp tribe, the cells are placed with their mouths downward, the nurses being enabled to attend to their charges by remaining on the floor of the next tier of cells. Taking one row of cells as an average, I counted twenty-four from the central aperture to the circumference, thus giving a tolerable notion of the number of cells in each tier. The aperture is not precisely in the middle, so that some rows of cells are necessarily larger than others, but I purposely selected a row which seemed to afford a fair average.

THERE are also certain British wasps which always make pensile nests, though none of them are so complicated or so finely constructed as those of the pasteboard wasps of hotter climates.

These are popularly called TREE WASPS, and the best known among these pensile wasps is the insect which is sometimes known as *Vespa Britannica*, but which is now named *Vespa Norwegica*, and may therefore be called the NORWEGIAN WASP.

Of the species in question Mr. Smith remarks that it is rare in the South and West of England, but is not uncommon in Yorkshire and plentiful in Scotland. It seems to be a nocturnal insect, for a collector of lepidoptera found that when 'sugaring' trees at night, for the purpose of attracting moths, numbers of these wasps settled on the sweet bait, and not only were more numerous than the lepidoptera, but actually resented any attempts at dislodgment.

The nest of this insect is always pensile, and is hung from the branches of a tree or shrub, the fir and gooseberry being the favourites. A pretty specimen in my own collection was taken from a gooseberry-tree in a garden, and another similar nest was found at no great distance. One of these nests I presented to the British Museum, and the other is now before me. It is very small, only having one 'terrace,' in which are thirteen cells, arranged in five rows, four being in the central row, and the rest graduating regularly. It is almost as large as a well-sized turnip radish, and something of the same shape, supposing the radish to be suspended by the root, and to be cut off just below the leaves. The outer envelope is composed of three layers overlapping each other, which are very fragile, considering the work they have to perform.

The wasp itself is prettily marked, and although it is variable in colouring, can be recognised by the black anchor-shaped mark on the clypeus, and the squared black spot on the segments of the abdomen.

ANOTHER species of British Tree Wasp is the CAMPANULAR WASP (*Vespa sylvestris*), a species which has received a multitude of scientific names, but which is not variable in colour as that which has just been mentioned. Though it has a wider distribution than the Norwegian Wasp, it is scarcely so plentiful an insect, and is remarkable for an occasional habit of making a subterranean nest like that of the common wasp. The NORTHERN WASP (*Vespa borealis* or *arborea*), is another of the pensile wasps, and is mostly found in the North of England and Scotland. Its nest is built in fir-trees. I may perhaps mention that the tree wasps may always be distinguished from their subterranean brethren by the colour of the antennæ, workers and females having the scape black in the ground wasps, and those which build in trees having it yellow in both sexes.

In the accompanying illustration are represented two nests, both from tropical America, and both found in similar localities. These are the habitations of two species of wasp, which are remarkable for their honey-making powers.

NECTARINIA. MYRAPETRA.

In the year 1780, a Spanish officer named Don Felix de Azara was raised from the rank of captain to that of lieutenant-colonel, and sent to Paraguay, in order to decide a dispute concerning the limits of the possessions respectively held by Spain and Portugal.

He was then thirty-four years of age, and being a man of great energy, set to work out the construction of a map of Paraguay. This was a Herculean task, occupying thirteen years in its completion, and forcing De Azara to explore regions before unknown, and to trust himself to the native tribes who had never before seen the face of a white man. While engaged in this occupation, he made a vast collection of notes upon the native tribes of Paraguay, as well as upon the beasts, birds, insects, and vegetation, together with an account of the method by which the Jesuit missionaries established themselves and ruled the country for many years.

After his return to Europe, in 1801, he published the account of his travels, and met with the usual fate of those who first penetrate into unknown countries. His statements were not believed, and among those which raised the greatest discredit was an account of certain wasps which made honey. Some persons said that the whole statement was a fabrication, and others remarked that the honey-making insects were simply bees which De Azara had erroneously considered to be wasps. Time, however, had its usual effect, and De Azara has been proved to be perfectly trustworthy in his remarks. The two specimens which are represented in the illustration are now in the British Museum, and afford tangible proofs that De Azara was right and his detractors wrong.

The right-hand figure represents the nest of a curious insect, named by Mr. Adam White *Myrapetra scutellaris.*

On looking at the exterior of the nest, our attention is at once excited by the material of which it is made, and the vast number of sharp tubercular projections which stud its surface. In colour it is dark, dull, blackish-brown, and its texture somewhat resembles very rough *papier-mâché*. On examining it with a pocket magnifier a matted structure is plainly visible, as if it were made of short vegetable fibres. This appearance accords with the accounts of the natives, who say that it is made from the dung of the capincha, one of the aquatic cavies of tropical America.

The whole of the exterior is thickly studded with projections, varying in size and shape, but being all of some sharpness at the tip. These projections are comparatively few at the top of the nest, becoming gradually more numerous as they approach the bottom, until at last they are set so thickly that the finger can scarcely be laid between them.

The object of these projections is not ascertained. The nest always hangs very low, seldom being more than three or four feet from the ground, and some writers say that the office of the sharp projections is to guard the nest from the attacks of the felidæ and other honey and grub-loving mammalia. Such may indeed be the true explanation, and indeed it is so obvious that no one could avoid seeing it. But I very much doubt whether a far better explanation is not in store, and I cannot see why the Myrapetra should stand in need of such protection, when the nest of the Nectarinia, which is placed in precisely the same conditions, is perfectly smooth and defenceless.

One use of the projections is evidently for the double purpose of concealing and protecting the entrance. On looking at the nest from above no entrance is visible, and it is not until after a close examination that the openings are found. They are concealed under a row of projections, which overhang them like the eaves of a house, and effectually keep off the rains which fall in such heavy torrents during tropical storms. The material of which these projections are made is the same as that of which the walls of the nest are built, except that it is very much thicker and harder, the various layers being hardly distinguishable, even with a good magnifier.

The interior of the nest is as remarkable as its exterior.

When cut open longitudinally, an operation which was carefully performed by Mr. White, a very curious sight presents itself. The nest is filled with combs, all very much curved, and these curves accommodating themselves beautifully to the general form of the nest. At the top is a nearly globular mass of brown paper-like substance, which is apparently the nucleus of the nest. The first comb closely surrounds this globular mass, leaving only a small interval between them, so that it forms part of a hollow sphere, and a section of it would present a form like that of the capital letter C laid on its back.

The rest of the combs follow in regular order, the curve of each becoming shallower, until the last is but slightly depressed in the centre. They are carried to the sides of the nest and thereto attached, except in a few places, where an open space is left between the edge of the comb and the side of the nest, so as to allow the wasps to have access to the different tiers of cells. As is the case with most of the wasp tribe, the tiers are single, and the mouths of the combs are all downwards.

The depth of the cells, and consequently the thickness of the combs, varies according to their position in the nest, the upper cells being the largest, and those below the smallest. The longest cells are from five to seven lines in length, and the shortest, about two lines. The material of which they are made is the same as that of which the exterior is formed, and is of quite as dark a colour. In texture, however, it is much slighter, being very thin and paper-like. These cells extend to the very edges of the combs, of which there are fourteen in the present specimen. The length of the nest is sixteen inches, and its diameter in the widest part is one foot.

In the upper combs was discovered a quantity of honey, which, when it was found, was hard and dry, of a deep brownish-red, and without either taste or scent. De Azara mentions that himself and some of his men ate the honey of the Myrapetra, and that it was of a deleterious character. Another species of honey-making wasp, *Polistes Licheguana*, a native of Brazil, was discovered by M. St. Hilaire, who mentions that it lays up in the nest a large provision of honey, which is very injurious to mankind, on account of the poisonous plants from which it is taken. *Polistes gallica* also fills its cells with honey, which, however, does not seem to be poisonous.

Within the nest were found also the remains of insects. There was the body of a black fly, which belongs or is allied to the genus *Bibio*, and the remains of a neuropterous insect, which apparently belongs to the genus *Hemerobius*.

The Myrapetra itself is of variable size, the largest being about four lines in length, and rather more than half an inch in expanse of wing. It is of a dusky brown colour, and is remarkable for having the first joint of the abdomen very much lengthened and narrowed, so that it sometimes resembles the same organ in the Pelopæus.

At the left hand of the same illustration may be seen a rather large globular nest, suspended from the boughs. This nest is shown in the position which it usually occupies, namely, hidden in the dark recesses of the Brazilian forest, amid the varied vegetation which grows so profusely in the hot and wet parts of the country which the insect frequents.

The name of the species which makes this nest is *Nectarinia analis*, a title which is significant and appropriate enough, but which is rather unfortunate, inasmuch as it has already been applied to a genus of birds, the well-known honey-suckers of Africa and India, which are so frequently mistaken for humming birds, on account of their small size, their brilliant plumage, their slender beaks, and their fondness for flowers.

This is not nearly so beautiful a nest as that which has just been described, the combs being devoid of regularity, and piled upon each other, as if the insect had no settled plan on which to work, and put each comb in any place where there happened to be room for it. Irregular, however, as the structure may seem, it is not without a kind of order, for though the combs look as if they had been placed in a heap, and then rolled together, so as to assume a partially spherical shape, they are at all events made with the intention of forming that shape, so that they may be included under a single covering. In the specimen in the British Museum, the outer wall of the nest has been broken away in several places, so as to permit the combs to be seen.

The entrance for the insects is very small, and when the respective dimensions of the wasp and the nest are taken into consideration, it seems really wonderful that when the inhabitants enter their house, they do not lose themselves in the intricate windings through which they pass from one comb to another.

The wasp which makes this nest is bee-like in form, and very small, not a quarter of an inch in length, and bearing some resemblance to those tiny solitary bees that are seen so plentifully upon dandelions and various umbelliferous flowers.

The nest is always hung near the ground, quite as low as that of the Myrapetra, and is suspended from the slender twigs and long, delicate leaves which are woven into its substance, and in many places pierce completely through the nest, and project through the outer covering. It is, however, destitute of the sharp projections which guard the home of the latter insect, and as the outer wall is both thin and fragile, it would fall an easy prey to any insect-eating animal that might take a fancy to it. I cannot but think that this utterly defenceless state of the Nectarinia's nest affords a proof that the spikes upon the habitation of the Myrapetra are not for the purpose of defending the nest against the attacks of enemies.

As is the case with the Myrapetra, the cells are made with walls much firmer than those of our English wasps or hornet, which are only intended to hold successive generations of young, and in consequence are made of a comparatively flimsy material, only strengthened very slightly at the entrance. Were honey to be placed in the cell of any known British wasp it would immediately soak into the walls of the cell, and thence escape by slow degrees, but as the young grub, which is the only tenant of the cell, is without feet and is not in the least formed foi locomotion, a very slight partition is sufficient to control its movements.

The grub does nothing but hold to the end of the cell with its piercers, open its mouth for food, and occasionally protrude or withdraw itself in a very slight degree; and its utter immobility in the larval and pupal states affords a strange contrast to the restless and fussy activity which actuates it after it has attained its perfect form.

A CREATURE is upon our list of pensile insects, which may also be reckoned among the social or parasitic insects, but which makes its habitation in such a manner that its proper place is among the pensiles. This is the pretty little ichneumon which is known to entomologists as *Microgaster alvearius*. The name Microgaster is of Greek origin and signifies 'little belly,' this being a very appropriate name for this insect, whose abdomen is of very small dimensions, and indeed appears to be just a little supplementary growth which might be removed without causing any inconvenience to the insect. It belongs to the same genus as a very common insect called *Microgaster glomeratus*, which will be duly described when the parasitic animals are under consideration.

With regard to this insect, I have been rather fortunate, having found many specimens of the nests, and bred from them several hundred insects.

Although plentiful enough in certain places, the BURNET ICHNEUMON, as I shall venture to call this species, is very local, and while abounding in one place may never be seen in another spot at the distance of a very few hundred yards. I give it the popular name of Burnet Ichneumon, for the same reason—comparing great things with small—that Caius Martius bore the title of Coriolanus and Publius Cornelius Scipio was termed Africanus—namely, that it destroys so many Burnet Moths.

In its perfect state the Ichneumon looks like a rather small gnat, and would probably be mistaken for that insect by a non-entomological observer. When examined through an ordinary magnifying glass, it is seen to possess a wondrous beauty which no one could ever suspect when looking at it with the unaided eye. The body and head are of a pale yellow colour, except the prominent compound eyes, which are dark blackish brown. The head is round and rather small, but the thorax is of enormous size, quite as proportionately large as the chest of a man would be did it project some eighteen inches in front and reach to his heels.

In singular contrast to the huge thorax is the very tiny abdomen, which is of a retort shape, curved, and fixed in the upper surface of the thorax by its smaller end. Indeed, the abdomen bears the same relation to the thorax, that the 'tick' in the capital letter Q does to the whole of the letter. The limbs are long, and, when the size of the insect is considered, are singularly powerful, especially the last pair of legs. We think the legs of the kangaroo are enormously large in proportion to the size of its body, but they must be doubled in length as well as in thickness to equal those of the Burnet Ichneumon. The fore-limbs are not so very large, but they are long and possessed of great clasping power, aided by the hooked feet.

What then is the use of such powerful limbs? The habits of the insect supply the answer.

As is the case with many ichneumon flies, this insect—which, by the way, is not a fly but a near relation to the bee and ant—deposits its eggs upon caterpillars, boring holes in their skin with its pointed ovipositor, which is the analogue of the bee's sting, and inserting its eggs in the perforations. As may naturally be imagined, the caterpillar has a very strong objection to this proceeding, and when the ichneumon settles upon it, and begins to use her weapon, twists and wriggles about like a captured eel.

Now the strong limbs of the ichneumon come into play. Minute as is the insect when compared to the caterpillar, bearing about the same relationship that a rabbit bears to an elephant, the legs are so long that they can include a considerable portion of the skin in their embrace, and so strong that they can retain their hold in spite of the contortions with which the caterpillar tries to rid itself of its persecutor. Retaining her place, therefore, the ichneumon deposits a great number of eggs in the poor caterpillar, and then goes to find another victim.

I am not sure whether or not the ichneumon makes a separate wound for every egg. If so, the feelings of the caterpillar are not to be envied, for I have found nearly a hundred and fifty ichneumon larvæ in the body of a single caterpillar. No wonder that the persecuted being endeavours to fling off the creature that is inflicting so many wounds. The numerous short and bristle-like hairs with which the legs are thickly clad, are doubtless useful in retaining the hold of the insect.

Our last example of the pensile nests formed by the hymenoptera is a truly remarkable one. For some time I could scarcely decide upon its place in the present work, whether it was to be ranked as an example of the pensiles, social insects, or builders. On account, however, of the locality which is chosen

NESTS OF POLISTES.

for it, and the peculiar method by which it is attached to the branch, I have decided upon placing it among the pensile nests.

Generally, the shape of the comb is nearly round, as is seen in the upper figure of the illustration. The cells are remarkable for their radiating form, the bases being a trifle smaller than the mouths, a peculiarity which would hardly be noticed in a single cell, but which produces the spreading outline when a number of them are massed together.

Some of the cells, those in the middle for example, are much longer than the others, and in the specimens in the British Museum many of them are closed at the mouth, showing that the insect is within, and has not yet attained its perfect state. Those on the circumference, however, are much shorter, and are entirely empty, not having been yet occupied. It is very possible that these cells would have been lengthened had the insects been left to themselves.

Although the circular shape is mostly the rule with these combs, so that they look something like withered dahlias or chrysanthemums, it is not the invariable form. If the reader will look at the lower figure in the illustration, he will see that it is much wider than long, and is apparently composed of two of the circular combs fixed together.

Now comes the curious part of the structure. The combs are not fastened directly to the branches, but are attached to footstalks which spring from their centre, and are firmly cemented upon the branch or twig. This group of cells is copied from the specimen in the British Museum, but ought to have been reversed, so that the mouths of the cells hang downwards. The observer should notice the wonderful manner in which the balance is preserved, the footstalk occupying as nearly as possible the centre of gravity.

The footstalks are made of the same *papier-mâché* like substance as the cells, only the layers are so tightly compressed together that they form a hard, solid mass, very much like the little pillars which support the different stories of an ordinary wasp's nest, but of much greater size. The position of the combs is extremely variable, some being nearly horizontal, and others perpendicular, as shown in the illustration. These nests came from Bareilly in the East Indies.

WE now come to the pensile lepidoptera, of which a number of specimens will be mentioned. They all belong to the moths, the pensile butterflies being content with suspending themselves by a couple of threads, without taking the trouble to build or spin a nest.

OIKETICUS AND ATLAS MOTH.

On the right hand of the accompanying illustration may be seen a large moth flying downwards, and just above it are a couple of oval objects attached to a slender bough. This moth is that magnificent insect the Atlas Moth (*Saturnia Atlas*), and the oval objects are the cocoons which are spun by its larva.

The Atlas Moth belongs to the same genus as the emperor moth of this country, and is a truly splendid insect, though without the beautiful colours which decorate the emperor. Creamy white, soft yellow, and pale brown are the chief tints of the Atlas Moth, but they are so beautifully blended, the plumage is of so downy a softness, and the expanse of wing is so great, that the Atlas holds its own even amid the more vividly coloured lepidoptera of its own country.

There are many members of this genus scattered over the different parts of the earth, the finest and largest specimens being found between the tropics In all the species the antennæ of the males are remarkable for their beauty, being deeply feathered, and shaped something like a spear-head with a triangular blade, and in many examples there is a loose membranous talc-like spot in the middle of the wing.

The cocoons of the Atlas Moth are made of silken thread, much like that of the common silkworm, the cocoon being large in proportion to the size of the moth, and the quantity of silk is necessarily very great. Although the thread is not so fine or glossy as that of the ordinary silkworm, it is strong, smooth, and serviceable, and capable of being woven into fabrics of much utility.

We now pass to the second insect represented in the illustration. This is the House-builder Moth (*Oiketicus Sandersii*), an insect which is common in many parts of the West Indies, in several places being so plentiful that the sight of its long pendent domiciles is anything but pleasant to the proprietor of a garden.

Out of five species of insects belonging to this singular genus, the present has been selected, because on the whole its habitation is more remarkable than that of any other species. Some of them make their nest in a much stiffer form than is depicted in the engraving, taking pieces of slender twigs and forming them into hollow cylinders, the twigs being laid parallel to each other, very much like the rods in the old Roman fasces, which were borne by the lictors before the consuls. So close indeed is the resemblance, that by some writers the insects have been called Lictor Moths.

The reader will observe that in the illustration the nest is shown as depending from the caterpillar, part of which protrudes from its mouth and the other part is hidden. This attitude is given because it is that in which the insect is generally seen. While young the caterpillar is so strong, and the house is so light, that it can carry the tail nearly upright.

Scraps of wood mixed with fragments of leaves are the materials which are used, and they are bound together very firmly by the silken threads with which so many caterpillars are endowed, whether they belong to the butterflies or moths. There is a tolerable degree of elasticity about it, especially at the mouth, which is slightly expanded so as to assume an irregular funnel-like shape, and can be drawn together at will by means of the silken threads attached to its circumference. The caterpillar has thus two means of guarding itself from attacks. If it is still clinging to a branch, it can retreat into the house and press the mouth so firmly against the branch that it is closed effectively, just as a limpet shelters its soft body by pressing the top of the shell against the rock. Or, if detached, it can pull the lips together and thus shut itself up in its strange house as completely as a box tortoise in its shell.

Not only does the creature reside in this nest during its larval condition, but also passes the pupal stage in it, and sometimes the whole of its life. As soon as it ceases from feeding, and is about to become a pupa, it retires far into its cell, shuts up the mouth, throws off its last caterpillar skin, and there remains until the larva has become a perfect insect. Should the moth be of the male sex, it creeps out of the domicile and speedily takes to wing, employing itself in the great object of its life, that of seeking a mate.

In ordinary cases, to find a mate seems to be no difficult task, but the House-builder Moth has no ordinary obstacles to overcome. The female never leaves her cell, for she would be more helpless as a moth than as a caterpillar. Among the British moths we have several species in which the females are wingless, but at all events they do look like moths which have been deprived of wings, and are able to move about with tolerable freedom. Of these wingless females, the common Vapourer moth (*Orgyia antiqua*), is a familiar example, its fat, rounded abdomen and little truncated rudiments of wings being known to all collectors.

But the female House-builder Moth is as utterly helpless a being as can well be conceived. She has not the least vestige of wings, and but the smallest indications of legs or antennæ. None but an entomologist would take her for a lepidopterous insect, or even for an insect at all, for she looks like a fat, down-covered grub, with very feeble limbs, which can scarcely support the body, and with antennæ that merely consist of a few rounded joints, entirely unlike the beautiful feathered plumes which decorate the male.

One of our commonest moths makes a really beautiful pensile nest, though it is hardly appreciated as it should be. I allude to the well-known Tiger Moth (*Arctia caja*), whose scarlet, white, and brown robes are so familiar to every one who cares for insects, or who happens to possess or take an interest in a garden.

In two of its stages the insect is very common. In the larval condition it is popularly known as the Woolly Bear, in consequence of the coating of long bristle-like hairs with which its body is profusely covered, and which project like the quills of a porcupine, or the spines of a hedgehog, whenever the creature rolls itself up, a movement which it always makes when alarmed. So elastic are the hairs, that the caterpillar may be thrown from a considerable height without suffering any injury, and in all probability their formidable appearance serves to deter foes from meddling with it.

When the caterpillar has ceased feeding, and is about to become a pupa, it ascends some convenient object, and then spins a beautiful cocoon, shaped very much like the grass hammocks made by the natives of tropical America, and bearing a considerable resemblance to them in general form, as well as in the loose and open meshes. So large, indeed, are the meshes made, that the inclosed insect can be seen through the network, from the time that the old wrinkled skin is cast off and pushed away in a heap by the white and shining chrysalis, to the time when the chrysalis shell is in its turn shattered, and the perfect moth creeps slowly into the air, all dull, and sodden, and bewildered, with its undeveloped wings looking like four mottled split peas rather than the beautiful members which they soon become, when the air has passed into their vessels, and their multitudinous folds have been shaken out.

AMONG the pensile insects may be reckoned the beautiful BURNET MOTH (*Anthrocera filipendulæ*), an insect which has already been mentioned, while treating of the pensile hymenoptera.

This insect, which is well known for its splendid colours of deep velvet green, and blazing scarlet, is also notable for the shape of its antennæ, which are so swollen towards the tips as to induce many persons to reckon the insect as a butterfly rather than a moth.

The shape of the cocoon of the Burnet Moth is not unlike that of the tiger moth, but its material and position are very different. The cocoon of the tiger moth is slung horizontally, in hammock fashion, while that of the Burnet is set perpendicularly, and fastened to the upper part of a grass stem, one side being firmly pressed against it. The substance of the cocoon is quite opaque, greyish, rather stout, very tough, and having the silken threads, of which it is chiefly made, so conspicuous, that many persons take the cocoon to be the work of a spider.

Sometimes in a field, or even in a limited portion of a field, these cocoons are so numerous that at a little distance they look almost as if they were the seeds of the plant rather than the cocoons of an insect. In such cases the moths themselves may generally be found near the cocoons, sometimes being on the ground and sometimes on the wing.

THE handsome OAK EGGER MOTH (*Gastropacha quercus*) affords another example of the pensile cocoon. Of these insects also I have had great numbers; and some specimens of the moth, chrysalis, and cocoon are now before me, the cocoon unchanged by the eighteen years which have elapsed since it was made, but the moth sadly faded, after the manner of its kind when exposed to the action of light.

Large as is the caterpillar of the Oak Egger moth, it is contracted into a comparatively small chrysalis when it assumes the pupal state, and makes a cocoon which only allows enough space for the pupa and the cast larval skin. The form of the cocoon is egg-shaped, whence the name of Oak Egger, and its substance is rather peculiar, being thin, hard, and rather brittle when quite dry. Externally it is surrounded by a loose layer of silken threads, by means of which it is attached to the plant on which it hangs; but the cocoon itself is smooth, very much the colour of half-charred paper, and in spite of its brittleness is possessed of some elasticity.

There is a smaller insect, popularly called the LITTLE EGGER MOTH (*Eriogaster lanestris*), which spins a cocoon of a similar structure, except that the walls are of even harder and more uniform texture, scarcely larger than a wren's egg, and of a substance which looks almost as if it were made of the same material as the egg. When broken, it is found to be even more brittle than that of the larger insect. Owing, in all probability, to the exceeding closeness of the structure, which would exclude air from the inhabitant, it is perforated with one or two very tiny and very circular holes, which look just as if some one had been trying to kill the insect by piercing the cocoon with a fine needle or pin.

Even from the outside these perforations are visible, but they are much more evident when the cocoon is opened. The object of these holes is, however, conjectural, and it would be a useful experiment to stop them with wax, in order to see whether the inclosed insect could be developed when the air was thus excluded. I believe that there are none of these holes in the cocoon of the large Oak Egger Moth, and if there be any such perforations, they are so minute as to escape notice.

We now pass to the enormous variety of caterpillars which are popularly called Leaf-rollers, because they make their homes in leaves which they curl up in various methods.

Some use a single leaf, and others employ two or more in the construction of their nests. Even the single-leaf insects display a wonderful variety in their modes of performing an apparently simple task. Some bend the leaf longitudinally, and merely fasten the two edges together, while others bend it transversely, fixing the point to the middle nervure. Some roll it longitudinally, so as to make a hollow cylinder corresponding with the entire length of the leaf, while others roll it transversely so that the cylinder is only as long as the leaf is wide, and a few species cut a slit in the leaf and roll up only a small portion of it.

The leaf-roller caterpillars belong to numerous species, and are plentiful enough, too plentiful indeed to please the gardener, who finds the leaves of his favourite trees curled up and permanently disfigured by these little marauders. All of them are of small size, and some so minute that the mere fact of their ability to roll up a leaf is something wonderful.

One of the most common among the Leaf-rollers is the pretty Oak Moth (*Tortrix viridana*). It is a little creature with four rather wide delicate wings, the upper pair of a soft leaf green, and the under pair of a greyish hue. In some seasons, the moths, or rather their larvæ, are so plentiful that great damage is done to the oak forests, tree after tree being so covered with them that scarcely a leaf escapes destruction, and the growth of the tree is consequently checked.

Like all Leaf-rollers, they feed on the green substance, or *parenchyma* of the leaf, and being ensconced within their tubular home can eat without fear of molestation. They are not very much afraid even of the small birds, for as soon as a bill is pushed into one end of the leafy cylinder, the caterpillar hastily 'bundles' out of the other—there is no other word which so fully expresses the peculiar action of the larva—and lowers itself towards the ground by a silken thread which proceeds from its mouth. In fact, it acts like a spider in similar circumstances.

Where these insects are plentiful, an absurd effect can be produced by tapping the branches of oak trees with a stick. As the stroke reverberates through the branch, the leaves, which appear to the casual passenger to be in their ordinary condition, give forth their inhabitants, and hundreds of tiny caterpillars descend in hot haste, each lowering itself by a thread and dropping in little jerks of an inch or two each. Some of them are more timid than the others, and descend nearly to the ground, but the general mass of them remains at about the same height. Another tap will cause them all to drop a foot or two lower, the stroke being felt even at the end of the suspending thread, and by administering a succession of such taps they will all be induced to come to the ground. There they will wait a considerable time, but presently one of them will begin to reascend, working its way upwards along the slender and scarcely visible line as easily as if it were crawling upon level ground. The least alarm will cause them to drop again, for they are then very timid, but if allowed to remain in peace, they speedily reach their cells and enter them with a haste that very much resembles the quick jerk with which a soldier-crab enters the shell from which he has been ejected.

If a tolerably smart breeze be blowing, the sight is still more curious, for the caterpillars are swung about through very large arcs, and, if the wind be steady, are all blown in one direction, so that their line forms quite a large angle with the level of the leaf to which the upper end is attached. The caterpillars, however, seem to be quite indifferent in the matter, and ascend steadily, whether the line be simply perpendicular, or whether it be violently blown about by the wind.

At the proper season of year, the moths are as plentiful as the larvæ, and a shake with the hand will cause a whole cloud of the green creatures to issue forth, producing a strangely confused effect to the eye as they flutter about with an uncertain and devious flight. A sweep with an ordinary entomological net will capture plenty of them, but in a few minutes they all disappear, some of them returning to the branches whence they had come, and others dropping to the ground. During the summer of 1864 they were very plentiful in Darenth Wood, the heavy growth of oaks giving them every encouragement.

THE insect which commits such devastation on the lilacs is generally the little chocolate-coloured moth called the LILAC MOTH (*Lazotænia ribeana*), though there are other allied species which infest the same plant. Anyone may see the damaged leaves for himself, and therefore I shall not particularly describe them, but pass at once to the mechanical powers which are involved in the task of curling the elastic leaf into cylindrical form.

Compare the size of the lilac leaf and of the newly hatched caterpillar, the latter being about as large as the capital letter I. That so minute a creature should roll up the leaf by main strength is of course an impossibility, and the method by which that consummation is attained is so remarkable an instance of practical mechanics that I must describe the operation at length.

If the reader will procure one of the rolled leaves, he will see that the cylindrical portion is retained in its place by a row of silken threads, which are individually weak, but collectively strong, holding the elastic leaf as firmly as Gulliver was held by the multitudinous cords with which he was fastened to the ground. That they should hold the cylinder in shape is to be expected, but the manner in which the cylinder is made is not so clear. The following is the process :—

First, the caterpillar attaches a number of threads to the point and upper edges of the leaf, and fastens the other ends to the middle of the leaf itself. It now proceeds to perform an operation which is precisely similar to the nautical method of 'bowsing' up a rope. In order to 'bowse' a rope taut, two men are employed, one of them pulling the nearly tightened rope at right angles so as to bend it, while the other continually belays it to the cleats. Now, the caterpillar performs precisely this operation, but without requiring the aid of an assistant, the 'bowsing' being performed by its feet, and the belaying by its spinneret. By thus hauling at, and tightening each line in succession, the caterpillar bends the leaf over slightly, and then attaches a fresh series of threads to keep it in its place. By repeating this process, and by continually adding fresh lines, the creature fairly bends the leaf into a hollow cylinder, and then crawls inside to enjoy its well-earned home.

I may here point out that the whole process of rolling the leaf affords an admirable example of mechanics as exhibited in nature, and that it is achieved by the well-known principle of exchanging space and time for power. Although the caterpillar cannot by any exertion of strength roll up the leaf in one minute, it is enabled to do so by dividing the work into a multitude of parts, and taking much longer time about it, just as a man who cannot lift a single weight of a thousand pounds may do so with ease by dividing it into ten parts, and in consequence, by taking up a considerable time in lifting the separate parts.

Again, in the silken bands which hold the rolled leaf in its place, we have an excellent example of accumulated power; neither of the threads being alone capable of enduring the tension, but their united strength being more than sufficient for the task. The threads themselves are exceedingly elastic, and by their combined force aid the caterpillar in rolling the leaf.

As soon as the caterpillar has entered its new home, it begins to feed, eating the green substance of the leaf, and generally leaving the nervures untouched. Sometimes the caterpillar lives for so short a time that a single leaf is sufficient for its subsistence; but there are some species which are obliged to repeat the task more than once.

THERE are other insects which also make their habitations in leaves; but, instead of rolling up the leaf and living inside the cylinder, they make their way between the two membranes, and there remain until they have undergone their transformation.

The reader must often have seen the leaves of garden plants and trees, especially those of the rose, traversed by pale winding marks, that look something like the rivers upon a map, and having mostly a narrow dark line running exactly along the middle. These curious marks are the tracks which are made by the various leaf-mining insects, while eating their way

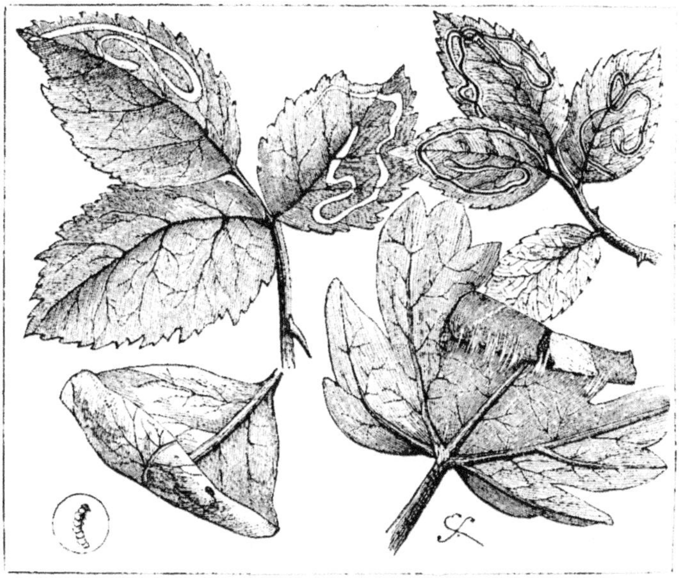

LEAF MINERS AND ROLLERS.

through the leaf in which they pass their larval state. In most cases, when the insect has completed its term of larval existence, one end of the track is found to be greatly widened, and to contain either the pupa itself or its empty case.

The track differs considerably in shape, according to the insect which makes it. Sometimes it winds about in the middle of the leaf, crossing itself more than once in its progress. Sometimes it proceeds in a nearly straight line across the leaf, and very frequently, especially in deeply-cut leaves, it follows the outline, keeping to the edge, and not trenching at all on the central portions.

Insects belonging to three orders are known to make these curious habitations; namely, the Lepidoptera, the Coleoptera, and the Diptera. Of these, the Lepidoptera are by far the most numerous, and belong to that group which is called, on account of their very minute dimensions, the Micro-Lepidoptera. These are all little moths, so small that on the wing they can scarcely be recognised as moths, and look more like little flies. They are all very beautiful, and many of the species are truly magnificent when seen through a microscope, their plumage glittering as if made of burnished gold and silver. Indeed, one genus in which these leaf-miners are comprised, is named Argyromiges, a title based on a Greek word signifying silver.

As for the beetle leaf-miners, they are to be found among the weevils; and it is a remarkable fact that one of these insects belongs to the genus Cionus, which in their larval condition are not only leaf-miners, but weavers of certain beautiful pensile cocoons.

Of the Diptera, the Celery Fly (*Tephritis onopordinis*) is a good example. The larva of this really pretty fly, with its green eyes and black and white spotted wings, feeds not only on the celery but on the parsnip, and does great harm to both plants. Gardeners often employ little boys to examine the celery plants, and whenever they find a 'blister,' as they technically call it, to crush the inclosed maggot between the fingers. The colour of this larva is pale green, so that it is not readily seen even when the blister is opened. If allowed to have its own way, the larva remains in the leaf until it has finished its eating, and then descends into the ground, where it changes into the pupal state, and remains until the following spring. In such a case, the leaves are often much damaged, the blisters being yellowish white, and the leaf itself drooping and half withered.

Our last examples of pensile nests are taken from the Arachnida, being formed by several species of spiders.

The best known of these creatures is the common GARDEN SPIDER (*Epeira diadema*), sometimes called the GEOMETRIC SPIDER, whose beautifully radiated net is so familiar that its general shape requires no description. Suffice it to say, that the spider exhibits wonderful skill in placing its web, making a framework of very strong threads or ropes, and then spinning the net itself between them. Very great strength is thus obtained, for the threads are exceedingly elastic; so that, although stretched tolerably tightly, they will yield to pressure, and immediately recover themselves. This property is very needful, in order to enable them to resist the wind, to which they are so fully exposed.

These spiders have, moreover, a most singular plan of strengthening their web, when the wind is more than ordinarily violent. If they find that the wind stretches their nets to a dangerous extent, they hang pieces of wood, or stone, or other substances to the web, so as to obtain the needful steadiness. I have seen a piece of wood which had been thus used by a Garden Spider, and which was some two inches in length and thicker than an ordinary drawing-pencil. The spider hauled it to a height of nearly five feet; and when by some accident the suspending thread was broken, the little creature immediately lowered itself to the ground, attached a fresh thread, ascended again to the web, and hauled the piece of wood after it.

It found this balance-weight at some distance from the web, and certainly must have dragged it for a distance of five feet along the ground before reaching the spot below the web. There were eight or ten similar webs in the same verandah, but only in the single instance was the net steadied by a weight.

The structure of the beautiful web is very remarkable.

It is nearly circular, and is composed of a number of straight lines, radiating from a common centre, and having a spiral line wound regularly upon them. Now, the structure of the radiating and the spiral lines is quite distinct, as may be seen by applying a microscope of moderate power. The radiating lines are smooth and not very elastic, whereas the spiral line is thickly studded with minute knobs, and is elastic to a wonderful degree, reminding the observer of a thread of India-rubber. So elastic, indeed, is this line, that many observers have thought that the spider has the power of retracting them within the spinnerets, inasmuch as she often will draw a thread out to a considerable length, and then, when she approaches the point to which it will be attached, it seems to re-enter the spinneret until it is shortened to the required length. This, however, is only an optical delusion, and caused by the great elasticity of the thread, which can accommodate itself to the space which it is required to cross.

One very remarkable point in the construction of these webs, so exactly true in all their proportions, is that they are executed entirely by the sense of touch. The eyes are situated on the front of the body and on the upper surface, whereas the spinnerets are placed at the very extremity of the body and on the under surface, the threads being always guided by one of the hind legs, as may be seen by watching a garden spider in the act of building or repairing her web. In order that the fact should be placed beyond a doubt, spiders have been confined in total darkness, and yet have spun webs which were as true and as perfect as those which are made in daylight.

A PECULIARLY beautiful pensile cocoon is constructed by a common British spider, scientifically termed *Agelena brunnea*, but which has no popular name.

The species whose beautiful nest will now be described is generally to be found upon commons, especially where gorse is abundant, as it generally hangs its nest to the prickly leaves of that shrub. The cocoon is shaped rather like a wine glass, and is always hung with the mouth downwards, being fastened by the stalk to a leaf or twig of the gorse. It is very small, only measuring a quarter of an inch in diameter, and when it is first made, is of the purest white, so as to be plainly visible among the leaves.

This purity, however, it retains but a very short time, for after the spider has deposited her eggs, which are quite spherical, and about forty or fifty in number, she closes the mouth of the cocoon and proceeds to daub it all over with mud. The moistened earth clings tightly to the silken cocoon, and disguises it so effectually that no one who had not seen it before that operation, could conceive how beautiful it had once been. The muddy cover makes the cocoon less visible, and may probably have another effect, that of protecting the inclosed eggs and young from the attacks of insects that feed upon spiders. Several other species have the habit of daubing their beautiful cocoons with mud.

This species is plentiful in Bostal Common and Bexley Heath in Kent, the profuse growth of gorse being very suitable to its mode of life, and I have several specimens of their nests taken from Shooter's Hill. June is the best month for them, as they may be found both before and after the mud has been applied.

An allied species, *Agelena labyrinthica*, is equally plentiful in similar localities, where its curious webs may be seen stretched in horizontal sheets over the gorse, and having attached to each web a cylindrical tube, at the end of which sits the spider itself. Heath and common grass are also frequented by this spider.

Besides the net or web in which it lives, and by means of which it catches prey, it makes a beautiful cocoon in which the eggs are placed. Externally the cocoon looks like a simple silken bag, perfectly white in colour, and, except in size, somewhat resembling that of the preceding species. It is only when quite freshly made, that the white hue of the cocoon is visible; for after its completion, it is covered with scraps of dry leaves, bark, earth, and other substances. If, however, this cocoon be opened, it is found to contain at least another cocoon within, and often comprises two, of a saucer-like shape, and made also of white silk. These inner cocoons are nearly half an inch in diameter, and contain a very variable quantity of pale yellow, spherical eggs, sometimes fifty in number, but often exceeding a hundred. The inner cocoons are firmly tied by strong lines to the interior of the large sac in which they are inclosed.

CHAPTER XV.

BUILDERS.

Building Mammalia—Definition of the title—Inferiority of the mammalia as architects—The BRUSH-TAILED BETTONG—Its structure and colour—The Nest of the Bettong, and its adaptation to the locality—Singular method of conveying materials—Its nocturnal habits—The MUSQUASH or ONDATRA—Its general habits—Its burrowing powers, and extent of its tunnels—The Musquash as a builder—Form and size of its house—Mode of killing the animal.

WE now take our leave of the Pensiles, and pass to those animals which build, rather than burrow or weave. The materials used by the Builders are variable. In the most perfect examples, earth is the material that is employed, but in many instances other substances such as wood, earth, and sticks are used by the architect.

As a general rule the mammalia are by no means notable for their skill in the construction of their houses. In making burrows they far excel all the other vertebrates both in the length of the tunnels and in the elaborate arrangement of the subterranean domicile. The mole, for example, is pre-eminent as a burrower and as a subterranean architect, and there are many of the rodents which drive a whole labyrinth of tunnels through the soil. But they are very indifferent builders, and with a few exceptions are unable to raise an edifice of any kind, or to weave a nest that deserves the name.

THE first example of the Building Mammalia is the PENCILLED BETTONG (*Bettongia pencillata*), sometimes called the BRUSH-TAILED BETTONG, and often known by the name of JERBOA KANGAROO. The word Bettong is a native name for a group of small kangaroos that are easily recognised by the shape of their heads, which are peculiarly short, thick, and round, and very unlike the long deer-like head of the larger kangaroos.

The Brush-tailed Bettong is about as large as a hare, and its tail is not quite a foot in length, though it appears longer in consequence of a brush-like tuft of long hair which decorates the end. It is a pretty creature, elegant in shape, extremely active, and the white pencillings on the brown back, the grey-white belly, and the jetty tuft on the tail are in beautiful contrast to each other.

The home of this animal is a kind of compromise between a burrow and a house, being partly sunk below the surface of the ground and partly built above it. The localities wherein the Bettong is found are large grassy hills whereon there is hardly any cover, and where the presence of a nest large enough to contain the animal, and yet small enough to escape observation, appears to be almost impossible. The Bettong, however, sets about its task by examining the ground until it finds a moderately deep depression, if possible near a high tuft of grass.

Using this depression as the foundation of the nest, it builds a roof over it with leaves, grass, and similar materials, not high enough to overtop the neighbouring herbage, and being very similar to it in external appearance. Grass of a suitable length cannot always be obtained close to the nest, and the Bettong is therefore obliged to convey it from a distance. This task it performs in a manner so curious, that were it not related by so accurate and trustworthy an observer as Mr. Gould, it could hardly be credited. After the animal has procured a moderately large bunch of grass, it rolls its tail round it so as to form it into a sheaf, and then jumps away to its nest, carrying the bunch of grass in its tail. In Mr. Gould's work on the Macropidæ of Australia, there is an illustration which represents the Bettong leaping over the ground with its grass sheaf behind it. After the nest has been completed, the mother Bettong is always careful to close the entrance whenever she leaves her home, pulling a loose tuft of grass over the aperture.

To an ordinary European eye, the homes of the Bettong are quite undistinguishable from the surrounding grass. The natives, however, seldom pass a nest without seeing it, and destroying the inmate. Being a nocturnal animal, the Bettong is sure to be at home and asleep during the daytime, so that when a native passes a nest he always dashes his tomahawk into its midst, thus killing or stunning the sleeping inmates.

OUR second and last example of the Building Mammalia is the MUSQUASH, or ONDATRA of North America (*Fiber Zibethicus*), sometimes called the Musk Rat.

This animal might have been placed among the burrowers, for it is quite as good an excavator as many which have been described under that title, but as it builds as well as burrows, it has been reserved for its present position in the work.

Essentially a bank-haunting animal, it is never to be seen at any great distance from water, and like the beaver, to which it is closely allied, it is usually to be found either in the river itself or on its edge, where its brown, wet fur harmonizes so well with the brown, wet mud, that the creature can scarcely be distinguished from the surrounding soil. It is seen to the best advantage in the water, where it swims and dives with consummate ease, aided greatly by the webs which connect the hinder toes.

The Musquash drives a large series of tunnels into the bank, excavated in various directions, and having several entrances, all of which open under the surface of the water. The tunnels are of considerable length, some being as much as fifty or sixty feet in length, and they all slope slightly upwards, uniting in a single chamber in which is the couch of the inhabitants. If the animal happens to live upon a marshy and uniformly wet soil, it becomes a builder, and erects houses so large that they look like small haycocks. Sometimes these houses are from three to four feet in height.

The natives take advantage of the habits of the animal, and kill it while it lies on its couch, much after the same manner as is used by the natives of Australia when they pass the house of the Bettong. Taking in his hand a large four-barbed spear, shaped something like the well-known 'grains' with which sailors kill dolphins and porpoises, the native steals up to the house, and driving his formidable weapon through the walls, is sure to transfix the inhabitants. Holding the spear firmly with one hand, with the other he takes his tomahawk from his belt, dashes the house to pieces, and secures the unfortunate animals.

In its subterranean home the Musquash lays up large stores of provisions, and in the habitation have been found turnips, parsnips, carrots, and even maize. All the roots had been dug out of the soil, and the maize had been bitten off close to the ground. The Musquash is not a large animal, the length of its head and body being only fourteen inches.

CHAPTER XVI.

BUILDING BIRDS.

The OVEN BIRD and its place in ornithology—Its general habits—Nest of the Oven Bird—Curious materials and historical parallel—The specimens in the British Museum—The internal architecture of the nest—Division into chambers—The PIED GRALLINA—The specimens at the Zoological Gardens—Materials and form of the nest—Boldness of the Bird—The SONG THRUSH and its nest—The BLACKBIRD and its clay-lined nests—Supposed reasons for the lining—The FAIRY MARTIN—Locality, shape, and materials of the nest—Social habits of the bird—How the nest is built—The HOUSE MARTIN—Material of its nest—Favourite localities—Ingenuity of the Martin—Adaptation to circumstances—The SWALLOW—Distinction between its nest and that of the Martin—Why called the Chimney Swallow.

AMONG the building birds, there is one species which is pre-eminently chief. Not only is there no equal, but there is no second. This is the OVEN BIRD (*Furnarius fuliginosus*), which derives its popular name from the shape and material of its nest.

The Oven Bird belongs to the family of the Certhidæ, and is therefore allied to the well-known Creeper of our own country. about as large as a lark, and is a bold-looking bird, rather slenderly built, and standing very upright. Its colour is warm brown. It is very active, running and walking very fast, and is much on the wing, though its flights are not of long duration, consisting chiefly of short flittings from bush to bush in search of insects. It generally haunts the banks of South American rivers, and is a fearless little bird, not being alarmed even at the presence of man. The male has a hard shrill note, and the female has a cry of somewhat similar sound, but much weaker.

The chief interest of this bird centres in its nest, which is a truly remarkable example of bird architecture. The material of which it is made is principally mud or clay obtained from the river banks, but it is strengthened and stiffened by the admixture of grass, vegetable fibres, and stems of various plants. The heat of the sun is sufficient to harden it, and when it has been thoroughly dried, it is so strong that it seems more like the

OVEN BIRD.

handiwork of some novice at pottery than a veritable nest constructed by a bird, the fierce heat of the tropical sun baking the clay nearly as hard as brick.

The ordinary shape of the nest may be seen by reference to the illustration, which is drawn from a remarkably fine specimen in the British Museum. It is domed, rounded, and has the entrance in the side. Its walls are fully an inch in thickness, and it seems strong enough to bear rolling about on the ground. This specimen was placed on a branch, but the bird is not very particular as to the locality of its nest, sometimes building it on a branch of a tree, sometimes on a beam in an outhouse, and now and then on the top of palings ; generally, however, it is built in the bushes, but without any attempt at concealment. Owing to its dimensions and shape, the nest is extremely conspicuous, and the utter indifference of the bird on this subject is not the least curious part of its history.

Strong as is the nest, it is still further strengthened by a peculiarity in the architecture, which is not visible from the exterior. If one of the nests be carefully divided, the observer will see that the interior is even more singular than the outside. Crossing the nest from side to side is a wall or partition, made of the same materials as the outer shell, and reaching nearly to the top of the dome, thus dividing the nest into two chambers, and having also the effect of strengthening the whole structure. The inner chamber is devoted to the work of incubation, and within it is a soft bed of feathers on which the eggs are placed. The female sits upon them in this dark chamber, and the outer room is probably used by her mate. The reader will remember that several instances of such supplementary nests have already been mentioned. The eggs are generally four in number.

Both sexes work at the construction of the nest, and seem to find the labour rather long and severe, as they are continually employed in fetching clay, grass, and other materials, or in working them together with their bills. While thus employed they are very jealous of the presence of other birds, and drive them away fiercely, screaming shrilly as they attack the intruder.

AUSTRALIA produces the two remarkable birds whose nests are given in the accompanying illustration.

The first of these feathered builders is the PIED GRALLINA (*Grallina Australis*), a bird which has become familiar to the public since its introduction to the Zoological Gardens. A pair of these birds have lived for some time in the Aquarium House, and have always attracted much attention as they fly to and fro in the large inclosure which is dedicated to them, to the dab-

FAIRY MARTIN. PIED GRALLINA.

chicks, kingfishers, wagtails, and other water-loving birds. Owing to the bold contrasts of black and white in their colouring they are very conspicuous, and their restless movements always attract the eye.

Although in its shape the nest of the Pied Grallina does not resemble that of the Oven Bird, the materials with which it is constructed are almost identical, consisting of mud and clay, in which are interwoven certain sticks, grasses, feathers, and stems of plants, which serve to bind the clay together, just as cow's hair binds together the plaster on our walls.

Like the Oven Bird, the Pied Grallina makes no attempt to conceal its nest, but places it quite conspicuously on a branch, as is shown in the illustration. It is almost invariably built on a bough which overhangs the water, and in spite of its weight and size, is fixed so firmly to the branch that there is no fear lest it should overbalance itself. The walls of the nest are very thick and solid, and the whole edifice looks very like an exceedingly rude and ill-baked earthenware vessel, just such an one, indeed, as Robinson Crusoe manufactured on his island. The bird is widely spread over Australia, so that its nest may be found in many parts of the country.

I MAY here mention that two of our best known song-birds form a basin-like nest of somewhat similar materials. Every one who has taken the nest of a SONG THRUSH (*Turdus musicus*), will remember that its interior is lined with a cup of a substance that resembles clay, but which is in fact composed chiefly of cowdung and decayed wood. This cup is exceedingly thin, but it is very hard and tough, and is so compact in its structure that it will hold water for some time. Like the mud wall of the Pied Grallina, it is strengthened by sticks and grass, with this difference, that whereas the latter bird incorporates the sticks and straws with the mud, the Thrush works the cup upon the sticks and straws.

The BLACKBIRD (*Turdus merula*), too, has a similar habit, only it employs veritable mud for the purpose, and spreads it in a much thicker layer than the Thrush. The eggs, however, are not placed on the dried mud, but on a layer of very fine grass. The object of this curious lining seems to be still undiscovered. Both the birds build in similar localities, and both make their nests close to the ground. It is possible that the stout walls may prevent the weasel or stoat from tearing the nest away from below, and so catching the young birds, but this is mere conjecture. Even the muddy lining does not repel all such attacks, for I once knew a dog that was in the habit of searching for nests of both these birds, and of eating the eggs and the young. He always obtained his prey by getting under the nest, biting out the bottom, and receiving the contents in his mouth.

THE curious flask-shaped nests which are seen in the illustration are built wholly of clay and mud, and are made by a beautiful little Australian bird, named the FAIRY MARTIN (*Hirundo Ariel*), closely allied, as its generic name signifies, to the swallows and martins of our own country. The bird is spread over the whole of Southern Australia, where it arrives in August, and whither it departs in September.

These remarkable nests are generally to be found upon rocks, and are always close to rivers, but have never been seen within many miles of the sea. Sometimes, however, the bird chooses another locality, and, instead of fixing its nests to the side of a rock, attaches them to the interior of one of the huge hollow trees which are so common in Australia. Now and then it behaves like the martin of England, and builds its nest under the protection of human habitations.

The shape of the nests always resembles that of a flask or retort, and their size is extremely variable, the length of the spouts, or necks, being from seven to ten inches, and the diameter of the bulb varying from four to seven inches. Mr. Gould mentions, in his work on the Birds of Australia, that each nest is the joint work of several birds, six or seven being sometimes employed upon one nest, one sitting in the interior, as chief architect, arranging and smoothing the material, while the others go off in search of mud and clay, which they knead well in their mouths before applying it to the nest.

As is generally the case with clay which is thus kneaded, it becomes very hard when baked in the sun, but, at the same time, is rather slow in drying. When the weather is dry, the bird can only work in the mornings and evenings, because the heat of the sunbeams soon renders the clay too stiff to be worked by the delicate beaks of the birds; and, therefore, in the middle of the day, the Fairy Martins cease from their architectural labours, and do nothing but chase flies. During wet weather, however, when no flies are abroad, and the air is full of moisture, the birds work continually at their nests, and soon complete their labours.

The exterior of the nest is quite as rough as that of the common English Martin; but in the interior it is beautifully smooth. The birds do not seem to have any particular care about the point of the compass towards which the entrance looks, but arrange it indifferently in any direction.

The Fairy Martin is a prolific little bird, laying four or five eggs, and rearing two broods in a year.

WE have several builders among our British birds, the best known of which is the common HOUSE MARTIN (*Chelidon urbica*), whose nests are so plentiful upon the walls of our houses.

The material of which the nests are built is a kind of mud, which becomes tolerably hard when dry, and is strong enough to exist for a series of years, and to serve for the bringing up of many successive broods. The bird is exceedingly capricious as to the spot which it selects for its residence, some houses being crowded with the mud-built nests, while others are free from them. The points of the compass are always noted by the Martin, for there are some points which it clearly detests, while it is equally fond of others. A wall with a north-eastern aspect is a favourite locality, while a southern wall is seldom chosen, probably because the heat of the meridian sun might dry the mud too quickly, or might cause inconvenience to the young birds.

My own house, however, forms an exception to this general rule, for the Martins have chosen to build on the south wall only, probably because the eaves project so far that after nine A.M. the nests are in shadow. Moreover, there is a narrow ledge, barely an inch in width, which runs under the eaves, and forms a support for the nests. While the Martins were engaged in bringing up their young, I ascended to the nests, and inspected them carefully, much to the indignation of the parent birds, who flew about wildly, darting occasionally out of their nests, and then stopping short and dashing away over the house. The opening of the nest being close against the eaves, the interior could not be inspected; but the touch of the finger showed that the walls were tolerably smooth, forming a great contrast with the rough exterior. The young birds were quite as much alarmed as their parents, and shrank to the very bottom of the nest, where they were quite invisible.

As to the nests themselves, they are exceedingly irregular on the outside, and look as if they had been made of that preternaturally ugly substance called 'rough-cast,' with which the walls of houses are sometimes disfigured. The material of which the Martin makes its nests is said to be the earth that is ejected by worms; but that this substance does not form the whole of the material is evident from the fact that stones, grass, and feathers are mixed with the mud, together with small twigs and a few fine roots of an inch or two in length.

The Martin is a rather ingenious bird, and is always ready to take advantage of any circumstance which may aid it in building its nest. The inch-wide ledge, for example, which I have just mentioned, has been quite appropriated by Martins, and there is scarcely a part of it which does not bear marks of their labours. At least a dozen nests have been begun and abandoned after a few beakfuls of mud have been put together, probably because the position is so exceedingly advantageous that the birds can scarcely begin in one place without regretting that they have not chosen a neighbouring spot.

THE common SWALLOW (*Hirundo rustica*) also makes a clay-built nest, similar in many respects to that of the martin, but differing in its shape. The nest of the martin is always covered, and entered by an aperture on one side. Mostly it is built immediately under a projecting ledge, which answers the purpose of a roof, but if no such accommodation can be obtained, it covers in the nest with a dome-like roof. The nest of the

Swallow, on the contrary, is open at the top, probably because the long forked tail would be crushed if pressed into so small a compass, while the shorter and simpler tail of the martin does not require so much space.

Wherever it can find an old chimney, the Swallow will always build its nest therein, a habit which has gained for the bird the popular title of Chimney Swallow. It will, however, build in many other situations, such as precipitous rocks and quarries, barns, outhouses, and steeples. There are usually five eggs, and the nest is lined with a soft bed of feathers, like that of the martin.

CHAPTER XVII.

BURROWING BIRDS—(continued)

Nesting of the Hornbills—Dr. Livingstone's account of the KORWÉ, or RED-BREASTED HORNBILL—The LONG-TAILED TITMOUSE—Its general habits—Its use to the gardener—Number of the young—Form and materials of the nest—Localities chosen by the bird—How to prepare the fragile eggs—The MAGPIE—Its domed and fortified nest—The common WREN and its nest—The LYRE BIRD—Origin of its name—Its domed nest—The ALBERT'S LYRE BIRD and its habits—The BOWER BIRD—Why so called—Civilisation and social amusement—The remarkable bower—Its materials and mode of construction—Use to which it is put—The Bower Birds in the Zoological Gardens, and their habits—Love of ornament—Meaning of the scientific name.

Two groups of large-billed birds are remarkable for their habit of nesting in hollow trees, and plastering up the entrance during the time of incubation. These are the TOUCANS of America and the HORNBILLS of Africa. We will take the latter birds as samples. The following interesting account of the Hornbill and its nest is quoted from Dr. Livingstone's well-known work.

'We passed through large tracts of Mopane country, and my men caught a great many of the birds called KORWÉ (*Tockus erythrorhynchus*) in their hiding-places, which were in holes in the mopane-tree. On the 19th (February) we passed the nest of a Korwé, just ready for the female to enter; the orifice was plastered on both sides, but a space was left of a heart shape, and exactly the size of the bird's body. The hole in the tree was in every case found to be prolonged some distance upwards above the opening, and thither the Korwé always fled to escape being caught. In another nest we found that one white egg, much like that of the pigeon, was laid, and the bird dropped another when captured. She had four besides in the ovarium.

'The first time that I saw this bird was at Kolobeng, where

I had gone to the forest for some timber. Standing by a tree, a native looked behind me, and exclaimed, 'There is the nest of a Korwé.' I saw a slit, only about half an inch wide and three or four inches long, in a slight hollow of the tree. Thinking the word 'Korwé' denoted some small animal, I waited with interest to see what he would extract; he broke the clay which surrounded the slit, put his arm into the hole, and brought out a *Tockus*, or Red-breasted Hornbill, which he killed.

'He informed me that when the female enters her nest, she submits to a real confinement. The male plasters up the entrance, leaving only a narrow slit by which to feed his mate, and which exactly suits the form of his beak. The female makes a nest of her own feathers, lays her eggs, hatches them, and remains with the young till they are fully fledged. During all this time, which is stated to be two or three months, the male continues to feed her and the young family. The prisoner generally becomes fat, and is esteemed a very dainty morsel by the natives, while the poor slave of a husband gets so lean that, on the sudden lowering of the temperature, which sometimes happens after a fall of rain, he is benumbed, falls down, and dies. I never had an opportunity of ascertaining the exact length of the confinement, but on passing the same tree at Kolobeng about eight days afterwards, the hole was plastered up again, as if in the short time that had elapsed the disconsolate husband had secured another wife. We did not disturb her, and my duties prevented me from returning to the spot.

'This (February) is the month in which the female enters the nest. We had seen one of these, as before mentioned, with the plastering not quite finished; we saw many completed, and we received here the very same account that we did at Kolobeng, that the bird comes forth when the young are fully fledged, at the period when the corn is ripe; indeed, her appearance abroad with her young, is one of the signs they have for knowing when it ought to be so. As that is about the end of April, the time is between two and three months. She is said sometimes to hatch two eggs, and when the young of these are full-fledged, other two are just out of the egg-shells: she then leaves the nest with the two elder, the orifice is again plastered up, and both male and female attend to the wants of the young which are left.'

PASSING from the birds which build with mud, we now come to those which use vegetable substances in their habitations. As examples of such architecture, we shall select the nests of those birds which are able to construct domed habitations, as well as the remarkable structures which are built by the Bower birds of Australia.

THE LONG-TAILED TITMOUSE (*Parus caudatus*) constructs a nest which is quite as wonderful in its way as the pensile home of the harvest mouse.

This pretty little bird is very plentiful in England, and owing to its habit of associating in little flocks of ten or twelve in number, and the exceeding restlessness of its character, is very familiar to all observers of nature. These flocks generally consist of the parents and their offspring, for the little creature is exceedingly prolific, laying a vast quantity of tiny eggs in its warm nest, and rearing most of the young to maturity. This is a bird which ought to be cherished by all possessors of fields or gardens, for there is scarcely a more determined enemy to the many noxious insects which destroy the fruits, vegetables, and flowers. Fortunately for ourselves, the Long-tailed Titmouse is very fond of the various saw-flies, that work such mischief among our fruit trees, and often lay waste whole acres of gooseberries, and it is no exaggeration to say that to a possessor of an orchard, or a fruit garden of any kind, every Long-tailed Titmouse is well worth its little weight in gold.

Although almost every one who lives in the country or who possesses a tolerably large garden in a town is perfectly familiar with this bird, comparatively few are in a position to narrate from personal observation the benefits which it confers upon us. The reason is simple; they do not rise early enough. A Long-tailed Titmouse in early morning, and the identical bird at noon, scarcely seem to be the same creature, so different are its ways.

It is a specially early bird, earlier than the sparrow, which is apt to be rather a sluggard as regards leaving its nest, though it sets up its garrulous chirp soon after daybreak. At that hour of the morning the Long-tailed Titmouse seems to cast off fear and diffidence, and allows itself to be watched without display-

THE LONG-TAILED TITMOUSE.

ing much alarm. Indeed, with the aid of a good opera-glass, it may be observed almost as well as if it were in a cage.

As the sun ascends above the horizon, and men and boys begin to go about to their daily work, the Titmouse loses its easy confidence, and will not suffer itself to be approached so closely as in the early morning. Generally, somewhere about five or six A.M. it leaves the garden and flies afield, and must then be sought far from human habitation. If, however, the garden should happen to be surrounded by walls, and the owner should happen to understand humanity as well as self-interest, the little bird will know that it will not be disturbed, and will remain in its sanctuary throughout the greater part of the day.

The quick, lively movements of the little creature are quite indescribable, so incessant and so varied are its changes of attitude. As it runs about the branches, it seems almost independent of gravity, and is equally at its ease whether its head, back, or breast be upward. It ever and anon utters an odd chirping note, which seems to issue from the bird as if it proceeded from some internal machinery, and were independent of the will of the creature which utters it. The observer should be careful to notice its quick, frequent pecks, and may be sure that every such movement denotes the slaughter of some insect, whether in the stage of egg, larva, pupa, or imago. The little beak is by no means so feeble as it seems, and is able to pick up an insect so small as would escape the observation of human eyes, or to pounce upon and destroy one which many a human being would not care to handle.

All the little flock, which are seen flitting about the trees, darting from branch to branch and tree to tree as if they were little arrows projected from bows, have at one time been inmates of the same nest, the beautiful domed structure which is shown in the illustration. How they are accommodated in so small a space seems quite a mystery, for not only is the hollow of the nest of no great size, but the interior is so filled with feathers and down that the space is still further limited.

The nest of the Long-tailed Titmouse is rather variable in shape, but its usual form is shown in the illustration. Generally, it is rather oval, and has an aperture at one side and near the top, through which the birds can pass. I believe that all domed nests, whether of bird or beast, are constructed by at least two architects, one of which remains within, while the other works from without. This is certainly the case with many creatures, and is probably so with all. The materials of which the nest is made are mosses of various kinds, wool, hair, and similar substances, woven by them with great firmness. It is remarkable that in the construction of this nest, which requires peculiar solidity, the Long-tailed Titmouse uses materials like those which are employed by the humming birds, and binds its nest together with the webs of spiders, and the silken hammocks of various caterpillars. The exterior of the nest is covered with lichens, so that the whole edifice looks very much like a natural excrescence upon the tree or bush in which it is placed, as is the case with the well-known nest of the chaffinch.

Sometimes the form of the nest is rather different from that which has been mentioned, and the structure is flask-shaped, the entrance corresponding to the neck of the flask. Now and then a nest is found in which there are two openings, one near the top in the usual position, and the other on the opposite side and near the bottom. The presence of one or two apertures is probably influenced by the position of the nest and the climate of the locality. If the finger be introduced into the aperture, a charmingly soft and warm bed of downy feathers is felt, *in* which, rather than *on* which, the numerous eggs repose.

The bird will build its nest in various trees, but always chooses a spot where the branches are very close and the foliage dense. The gorse bush is a favourite residence of the Long-tailed Titmouse, and so deeply is the nest buried in the prickly branches, that it cannot be removed without the aid of thick leather gloves, and a sharp, strong knife. Some skill and artistic taste are required in order to secure a good specimen, and it is difficult to hit the happy medium between cutting away too many branches, and retaining so many that the shape of the nest cannot be seen for their luxuriance.

The number of eggs is rather variable, but is always great, and on an average, some ten or twelve eggs can be found in a nest. They are so small and so fragile that the novice finds great difficulty in emptying them without breaking their delicate shells. This task may, however, be accomplished with perfect ease and safety if managed in the right way. Each egg should be enveloped in repeated wrappers of silver paper, soaked in a solution of gum arabic, one layer being allowed to dry before the next is added. When they are dry, a little hole is easily drilled on one side by means of a needle, the contents of the egg are then broken up with the same needle, and are washed out by injecting water through a very delicate glass tube. Anyone can make these slender tubes by merely taking a piece of ordinary glass tubing, heating it in a spirit lamp, and drawing the ends apart. It may then be broken off to form a tube of any degree of fineness, and by alternate injection of water and sucking the diluted contents into the tube, the egg will soon be emptied. The paper is removed by soaking in warm water.

WE have another well-known bird, which makes a nest as well domed as that of the long-tailed titmouse, though not nearly so pretty nor so elegant. This is the common MAGPIE (*Pica caudata*).

The nest of the Magpie is of very large size when compared with the dimensions of the architect, probably on account of the long tail of the mother bird, which cannot be protruded over the edge of the nest, as is the case with many long-tailed birds. It is not merely made of moss and similar soft substances, but the framework is very strongly constructed of sticks, among which are generally interwoven a number of sharp thorns, so that the nest is nearly as unpleasant to the bare hand as a thistle. Moreover, the bird has a way of gathering the thorns round the entrance, so that the hand cannot be inserted into the nest without danger of many wounds. Indeed, the nest is so large, and the eggs lie so far from the entrance, that to extract them is generally a task that cannot be accomplished without the aid of a knife.

Besides the thorny defence, the nest is mostly strengthened by its very position, being generally fixed in the furcation of several stout boughs, so that it can only be approached in certain parts. Moreover, the great height at which the Magpie loves to build the nest renders the operation of robbing it so dangerous, that many a nest escapes because no one has nerve enough to risk the ascent.

The position of the nest, too, conceals its true form so well, that a very practised eye is needed to distinguish it from an ordinary swelling of the bough, or from the heaps of dislodged twigs which are so often found in the forked branches of trees.

ANOTHER of our feathered dome-builders is the common WREN (*Troglodytes vulgaris*). The form and colouring of this bird are too well known to need description, and we shall therefore pass at once to its mode of nesting.

The Wren is rather peculiar in its method of constructing the nest, for though it can build a dome when there is need for it, and generally does so, it does not always choose to take so much trouble, but contents itself with an open nest arched over by a natural dome. Wherever it can find a convenient cavity, it will make its nest therein, building either no dome at all, or one of very flimsy construction, and such nests can generally be found in the holes of ivy-covered walls, under eaves, or among the thickly growing branches of fir-trees.

During the time when the Wren is building its nest, its loud, cheerful voice is heard in full perfection, and so full and powerful are its tones that the tiny bird seems hardly able to produce them. It is but a short song, and is little varied, the bird repeating nearly the same melody time after time within a few minutes. The long-drawn song of the nightingale, or the mellow notes of the thrush, are beyond the power of the Wren, but there are few birds whose song is more enlivening, or which add so much to the pleasure of a country walk. Besides the more formal song, the Wren has a pretty little monosyllabic chirp, which it utters as it pops about the hedges with its peculiar movements, dropping and ascending again with restless activity. The bird is so bold, too, that it will perch on a branch or a paling within a yard or two of the observer, and pour forth its bright song without displaying the least alarm.

As to the materials of the nest, the bird is no way fastidious, and generally seems to regard quantity rather than quality. Grasses of various kinds usually form the bulk of the nest, together with mosses, lichens, and similar substances. Withered leaves are generally worked into the nest, and I have more than once found specimens which were almost wholly composed of leaves. The size of the nest is wonderfully large, when the dimensions of the tiny architect are taken into consideration, and however large may be the hole in which the Wren makes its nest, it is nearly filled with the mass of grass, leaves, and wool which the Wren has conveyed into it. The interior of the nest is always warmly lined, sometimes with feathers, and sometimes with hair, and in the lining are generally some six or eight little eggs, nearly white, and covered with very minute red specks.

As is the case with the redbreast and one or two of our more familiar birds, the Wren will sometimes enter houses and build its nest in curtains, on shelves, and similar localities, while the interior of a disused greenhouse or stable loft is nearly sure to be tenanted by a Wren and its little brood.

AUSTRALIA is proverbially a strange land, and it is only in Australia, or perhaps in Madagascar, that we should look for a wren measuring some seventeen inches in height. Such a bird is, however, to be found in Australia, and is known to the natives by the name of BULLEN-BULLEN, and to the Europeans as the LYRE BIRD (*Menura superba*). It is remarkable by the way that the genius of the Australian language causes many words to be doubled, so that the natives speak of a well-known Australian marsupial as the devil-devil, and of a domestic servant as Jacky-Jacky.

New South Wales is the chosen country of the Lyre Bird, which is rather local, and affects certain defined boundaries. Its native name is derived from its peculiar cry, and the popular European name is given to the bird on account of the shape of its tail feathers. The two exterior feathers are curved in such a manner, that when the whole tail is spread they exactly resemble the horns of an ancient lyre, the place of the strings being taken by a number of slender decomposed feathers which rise from the centre of the tail. When the bird is quietly at rest, the tail-feathers cross each other at the curves, and present a very elegant appearance, though not in the least resembling a lyre. In general shape the bird bears some resemblance to a small turkey, except that the legs are longer and more slender, and that the feet do not resemble those of a gallinaceous bird. It is rather remarkable that the egg presents as curious a mixture of the insessorial and gallinaceous aspects as the bird itself.

The nest of this bird is not at all unlike that of the wren, being very much of the same shape, and domed after a similar fashion. The nest is, however, a very rough piece of architecture, composed almost wholly of twigs, roots, and various sticks, which are interwoven in a very loose, but very ingenious manner, so as to form a structure of tolerable firmness, which can be lifted and even subjected to rough treatment without being broken. At first sight it looks like those heaps of dead twigs which are so common in the birch-tree, but a closer inspection shows that there is a certain regularity in the disposition of the sticks, and that the bird is not without method, though that method be not at first apparent.

OUR last example of the Building Birds will be the well-known BOWER BIRD of Australia (*Ptilonorhynchus holosericeus*).

Perhaps the whole range of ornithology does not produce a more singular phenomenon than the fact of a bird building a house merely for amusement, and decorating it with brilliant objects as if to mark its destination. Such a proceeding marks a great progress in civilisation, even among human races. The savage, pure and simple, has no notion of undergoing more labour than can be avoided, and thinks that setting his wives to build a hut is quite as much labour as he chooses to endure.

The native Australians have no places of amusement. They will certainly dance their corrobory in one part of the forest in preference to another, but merely because the spot happens to be suitable without the expenditure of manual labour. The Bushman has no place of resort, neither has the much farther advanced Zulu Kafir. Even the New Zealander, who is the most favourable example of a savage, does not erect a building merely for the purpose of amusement, and would perhaps fail to

THE BOWER BIRD.

comprehend that such an edifice could be needed. Such a task is left to the civilised races, and it is somewhat startling to find that in erecting a ball-room, or an assembly room, or any similar building, we have been long anticipated by a bird which was unknown until within the last few years. Truly, nothing *is* new under the sun.

The ball-room, or 'bower,' which this bird builds is a very remarkable erection. Its general form can be seen by reference to the illustration, but the method by which it is constructed can only be learned by watching the feathered architect at work. Fortunately there are several specimens of this bird at the Zoological Gardens, and I have often been much interested in seeing the bird engaged in its labours.

Whether it works smartly or not in its native land I cannot say, but it certainly does not hurry itself in this country. It begins by weaving a tolerably firm platform of small twigs, which looks as if the bird had been trying to make a door mat and had nearly succeeded. It then looks for some long and rather slender twigs, and pushes their bases into the platform, working them tightly into its substance, and giving them such an inward inclination that, when they are fixed at opposite sides of the platform, their tips cross each other, and form a simple arch. As these twigs are set along the platform on both sides the bird gradually makes an arched alley, extending variably both in length and height.

When the bower is completed, the reader may well ask the use to which it can be put. It is not a nest, and I believe that the real nest of this bird has not yet been discovered. It serves as an assembly-room, in which a number of birds take their amusement. Not only do the architects use it, but many birds of both sexes resort to it, and continually run through and round it, chasing one another in a very sportive fashion.

While they are thus amusing themselves, they utter a curious, deep, and rather resonant note. Indeed, my attention was first attracted to the living Bower Bird by this note. One day as I was passing the great aviary in the Zoological Gardens, I was startled by a note with which I was quite unacquainted, and which I thought must have issued from the mouth of a parrot. Presently, however, I saw a very glossy bird, of a deep purple hue, running about, and occasionally uttering the sound which had attracted me. Soon, it was evident that this was a Bower Bird engaged in building the assembly-room, and after a little while he became reconciled to my presence, and proceeded with his work. He went about it in a leisurely and reflective manner, taking plenty of time over his work, and disdaining to hurry himself.

First he would go off to the further end of the compartment, and there inspect a quantity of twigs which had been put there for his use. After contemplating them for some time, he would take up a twig and then drop it as if it were too hot to hold. Perhaps he would repeat this process six or seven times with the same twig, and then suddenly pounce on another, weigh it once or twice in his beak, and carry it off. When he reached the bower he still kept up his leisurely character, for he would perambulate the floor for some minutes, with the twig still in his beak, and then perhaps would lay it down, turn in another direction, and look as if he had forgotten about it. Sooner or later, however, the twig was fixed, and then he would run through the bower several times, utter his loud cry, and start off for another twig.

Ornament is also employed by the Bower Bird, both entrances of the bower being decorated with bright and shining objects. The bird is not in the least fastidious about the articles with which it decorates its bower, provided only that they shine and are conspicuous. Scraps of coloured ribbon, shells, bits of paper, teeth, bones, broken glass and china, feathers, and similar articles, are in great request, and such objects as a lady's thimble, a tobacco-pipe, and a tomahawk have been found near one of their bowers. Indeed, whenever the natives lose any small and tolerably portable object, they always search the bowers of the neighbourhood, and frequently find that the missing article is doing duty as decoration to the edifice.

This species is more plentiful than another Bower Bird which will presently be described. As is the case with many birds, the adult male is very different from the young male and the female in his colouring. His plumage is a rich, deep purple, so deep indeed as to appear black when the bird is standing in the shade. It is of a close texture, and glossy as if made of satin, presenting a lovely appearance when the bird runs about in the sunbeams. The specific name, *holosericeus*, is composed of two Greek words signifying all silken, and is very appropriate to the species. The female is not in the least like the male, her plumage being almost uniform olive green, and the young male is coloured in a similar manner.

CHAPTER XVIII.

BUILDING INSECTS.

The TERMITE, or WHITE ANT—General habits of the insect—African Termites and their homes—Termites as articles of food—American Termites—Mr. Bates account of their habits—European Termites—Their ravages in France and Spain—M. de Quatrefages and his history of the Termites of Rochefort and La Rochelle—The TRYPOXYLON of South America—The PELOPŒUS and its curious nest—Mr. Stone's Wasp nests and their history—Difference of material—The FORAGING ANTS of South America and their various species—Nests and habits of the Foraging Ants—The AGRICULTURAL ANT of Texas—Dr. Lincecum's accounts of its habits.

WE now pass to the many insects which may be classed among the Builders. The reader will probably notice that several of the true builders are omitted in this department, but will find them under the head of Social Insects.

OF the Building insects the TERMITE, or WHITE ANT, as it is popularly and wrongly called, is the acknowledged head and chief. There are certain other insects that erect habitations which are truly wonderful, but there is not one that approaches the Termite in the size of its building or the stone-like solidity of the structure.

The history of the Termites is so complicated, and so full of incident, that I might occupy several hundred pages of this work in describing them and their nests, and yet not have exhausted the subject. I shall, therefore, give a general sketch of the Termites and their habits, and then relate a few details concerning the species which are found in Africa, Asia, America, and Europe.

In the first place, the reader must understand that the Termite is not an ant at all, but belongs to a totally different order of insect, and is allied to the dragon-flies, the ant-lions, the May-flies, and the beautiful Lace-wing flies.

The Termites are social, and, like other social insects, are divided into several grades, such as workers, males, and females, the two latter of which are winged when they reach maturity. The body is oblong and flat, the antennæ short, and the mandibles flattened and toothed, and in most cases extremely long and formidable. Each colony is founded by a single pair, popularly called the king and queen, the rest of the population consisting of developed males and females, which are intended to perpetuate the species and found fresh colonies, and of undeveloped individuals, or neuters, of both sexes. The neuter males are termed soldiers, and are armed with powerful jaws proceeding from enormous heads, and the neuter females are termed workers, and are very small.

There are now before me some specimens of African Termites, the soldiers of which are five or six times as large as the workers. They are formidable creatures, but they can do little harm beyond inflicting a severe bite, as they are not furnished with stings nor even with poison glands. They can bite through the clothes of an European, and when they swarm upon the bare limbs of the negro, they inflict almost unbearable tortures. The chief duty of the soldier seems to be the defence of the nest; for whenever the walls are broken down the soldiers come trooping out to attack the invader, and being quite unconscious of fear, they will seize on the first strange object that happens to come in their way. There are comparatively few soldiers, their proportion to the workers being only one per cent.

When a pair of developed Termites have settled themselves to form a colony, they share the fate of certain Oriental potentates, and never move out of their royal cell. When the queen is fairly settled, she increases in size so rapidly, that, even if she were set at liberty, she could not crawl an inch. While the head, thorax, and legs retain their original dimensions, the abdomen swells until it is more than two inches long and about three quarters of an inch in width. Thus developed, she produces eggs by the thousand, which are immediately carried off by the workers, who have reserved certain apertures in the royal apartment through which they can easily pass. When the eggs are hatched, the young are carefully watched and tended until they are at last developed into males, females, or neuters, and themselves are able to take part in the manual work.

A full-sized nest of the African Termite is a wonderful structure. Although made merely of clay, the walls are nearly as hard as stone, and quite as hard as the brick of which 'villa residences' are usually built. The form of the nest is essentially conical, a large cone occupying the centre, and smaller cones being grouped round it, like pinnacles round a Gothic spire.

In Anderson's valuable work, 'Lake Ngami,' there are many detached accounts of the African Termite. He states that he has seen nests which were full twenty feet in height, and had a circumference of a hundred feet, and that when the insects were developed and obtained their wings, they issued forth in such hosts that the air seemed as if it were filled with dense and white snow-flakes. So strong is the instinct for rushing into the air, that they can scarcely be retained within the nest, and will even pass through fire in order to gain their end.

The nests are always interesting objects, even from the exterior. The walls are so hard that hunters are accustomed to mount upon them for the purpose of looking out for game, and the wild buffalo has a similar habit, the structure being strong enough even to support the weight of so large an animal. The daily labours of the architects can easily be traced, on account of the dampness of the recent clay, so that an approximation can be formed as to the length of time which is occupied in erecting one of the nests. The traveller is always glad to see a large Termite nest, because he is nearly sure to find the surface studded with mushrooms, which are larger and better flavoured than those which our fields produce.

The natives have another motive for looking after the Termite nests, because they eat the inmates, considering them to be a peculiar luxury. The same author whom I have already mentioned, describes a curious interview that he had with Palani, a

Bayeiye chief. Wishing to show the chief the superiority of European cookery, Mr. Anderson spread some apricot jam on bread, and offered it to him. The chief took it, and expressed himself much pleased with it, but asserted that Termites were much superior in flavour. In order to catch the Termites in sufficient numbers, the native makes a hole in the nest, and when the workers are congregated for the purpose of repairing the breach, he sweeps them into a vessel, and repeats the operation until he has obtained as many as he wants.

As is the case with the true ants, the Termites only retain their wings for a limited period, using them for the purpose of escaping from the nest, and snapping them off as soon as they have met with a partner. The manner in which the wings are fixed to the body is the same in both groups of insects, and these singular organs are shed by being bent sharply forwards. If a living Termite be caught, and its wings pressed forward with a pin, they will instantly snap off; but if bent backwards, a piece of the body will be torn away before the wings can be removed.

As to the Termites of Southern America, much information may be obtained from Mr. Bates's valuable work on the natural history of the Amazons. As many of his remarks simply prove the identity of habits between the Termites of the old world and those of the new, I shall say nothing about them, but merely give a brief abstract of his observations.

As with the species which have already been described, the soldiers are the only individuals that fight. When, therefore, the ant-bear tears down the walls of the nest and begins to lick up the inmates, none but the soldiers are killed, they having come out to fight the enemy, while the workers have all run away and hidden themselves underground. In consequence of this fact, the economy of the nest is but slightly disturbed, and after the ant-bear has gone away, the workers begin to raise their walls afresh.

It must be remembered that the nests of the Termite are not confined to the surface, but extend to a considerable distance in the earth, the subterranean galleries being proportionately large to the superimposed nest. Indeed, the greater part of the material with which the walls and galleries are built is brought from below and carried upwards through the nest itself. There is no visible outlet to a Termite's nest, because the insects construct long galleries through which they can pass without suffering inconvenience from the light of day. Both the workers and soldiers are blind; but, in spite of the absence of external visual organs, they are very sensitive to light, and avoid it in every possible way.

The food of the Termite is of a vegetable character, and consists mostly of wooden fibres. They will, however, eat through almost anything, and the traveller in hot climates finds them among his worst troubles. They will cut to pieces the mat on which a man is lying. They will eat nearly all the wood of his strong box, leaving a mere shell no thicker than the paper on which this account is printed. They will devour all his collection of plants, beasts, birds and insects; and a table or any other article of furniture, if left too long in one position, will be utterly ruined by the Termites, which have a fashion of eating away all the interior, but leaving just a thin shell, which looks as if nothing were the matter.

When the adult Termites leave their homes, they often fly in such clouds that they fill the rooms, and even put out the lamps by their numbers. As soon as they touch ground they shed their wings, and then they begin to find how many enemies they have. Of the myriad hosts that pour into the evening air, not one in twenty thousand survives to found a new colony. They have foes above, below, and on every side. The bats and goatsuckers hold high festival on these evenings when the Termites are abroad, and after the insects have cast their wings they are pursued by ants, toads, spiders, and a host of other enemies.

WE will now pass to the European Termites, whose history is elaborately given by M. de Quatrefages. Rochefort, Saintes, and Tournay-Charente have for some years suffered from the ravages of the Termites, and now La Rochelle is invaded by these terrible destroyers. In all probability they were imported by some ship, taken ashore in the boxes into which they had penetrated, and thence spread into the country around. Efforts are being made towards the extirpation of these terrible insects, but nothing seems as yet to have had any great effect. How serious are the damages which they work may be seen from the following account by M. de Quatrefages, in his 'Rambles of a Naturalist,' vol. ii. p. 346 :—

'The Prefecture and a few neighbouring houses are the principal scene of the destructive ravages of the Termites, but here they have taken complete possession of the premises. In the garden, not a stake can be put into the ground, and not a plank can be left on the beds, without being attacked within twenty-four hours. The fences put round the young trees are gnawed from the bottom, while the trees themselves are gutted to the very branches.

'Within the building itself, the apartments and offices are alike invaded. I saw upon the roof of a bedroom that had been recently repaired, galleries made by the Termites which looked like stalactites, and which had begun to show themselves the very day after the workmen had left the place. In the cellars I discovered similar galleries, which were within half-way between the ceiling and the floor, or running along the walls and extending no doubt up to the very garrets; for on the principal staircase other galleries were observed between the ground floor and the second floor, passing under the plaster wherever it was sufficiently thick for the purpose, and only coming to view at different points where the stones were on the surface; for, like other species, the Termites of La Rochelle always work under cover wherever it is possible for them to do so.

'MM. Milne-Edwards and Blanchard have seen galleries which descended without any extraneous support from the ceiling to the floor of a cellar. M. Bobe-Moreau cites several curious instances of this mode of construction. Thus, for instance, he saw isolated galleries or arcades, which were thrown horizontally forward like a tubular bridge, in order to reach a piece of paper that was wrapped round a bottle, the contents of a pot of honey, &c.

'It is generally only by incessant vigilance that we can trace the course of their devastations and prevent their ravages. At the time of M. Audoin's visit a curious proof was accidentally obtained of the mischief which this insect silently accomplishes. One day it was discovered that the archives of the Department were almost totally destroyed, and that without the slightest external trace of any damage. The Termites had reached the boxes in which these documents were preserved by mining the wainscoting; and they had then leisurely set to work to devour these administrative records, carefully respecting the upper sheets and the margin of each leaf, so that a box which was only a mass of rubbish, seemed to contain a pile of papers in perfect order.'

In the British Museum are several examples of the ravages worked by Termites, one of which is an ordinary beam that has been so completely hollowed and eaten away, that nothing remains but a mere shell no thicker than the wood of a bandbox.

Besides the species which were investigated by M. de Quatrefages, there are others in the south of France, and in Sardinia and Spain. One species, *Termes flavicollis*, chiefly attacks and destroys the olives, while in the Landes and Gironde the oaks and firs are killed by another species, *Termes lucifugus*.

As the limits of the work preclude a very lengthened account of any one creature, our history of the Termites must here be concluded, although much interesting matter remains unwritten.

In the accompanying illustration are figured the nests of two insects, both of them natives of tropical America, and both belonging to the hymenopterous order. The upper insect is known to entomologists by the name of *Trypoxylon aurifrons*, but has at present no popular name.

This insect makes a great number of earthen cells, shaped something like those of the last-mentioned species; the cells being remarkable for the form of the entrance, which is narrowed and rounded as shown in the figure. In some cases the neck is so very narrow in proportion to the size of the cell, and the rim is so neatly turned over, that the observer is irre-

TRYPOXYLON AND PELOPÆUS.

sistibly reminded of the neck of a glass bottle. The insect makes quite a number of these nests, sometimes fastening them to branches, as shown in the illustration, but as frequently fixing them to beams of houses. It has a great fancy for the corners of verandahs, and builds therein whole rows of cells, buzzing loudly the while, and attracting attention by the noise which it makes.

THE lower insect is the pretty *Pelopœus fistularis*, with its yellow and black banded body. Both the insects, as well as their houses, are represented of the natural size.

The cell of the Pelopæus is larger than that of the preceding insect, and occupies much more time in the construction, a week at least being devoted to the task. She sets to work very methodically, taking a long time in kneading the clay, which she rolls into little spherical pellets, and kneads for a minute or two before she leaves the ground. She then flies away with her load, and adds it to the nest, spreading the clay in a series of rings, like the courses of bricks in a circular chimney, so that the edifice soon assumes a rudely cylindrical form.

When she has nearly completed her task, she goes off in search of creatures wherewith to stock the nest, and to serve as food for the young, and selects about the most unpromising specimens that can be conceived. Like many other solitary hymenoptera, this Pelopæus stores her nest with spiders, and any one would suppose that she would choose the softest and the plumpest kinds for her young. It is found, however, that she acts precisely in the opposite manner.

THERE is in the British Museum a most extraordinary series of wasps' nests, built by the insects under the superintendence of the late Mr. Stone, whose death is a serious loss to all zoologists. The story of these nests is very remarkable, and shows how much we have to learn concerning the habits and instincts of insects.

In the month of August, 1862, a nest of the common WASP (*Vespa germanica*) was taken near Brighthampton, and handed over to Mr. Stone, who has long been in the habit of experimenting upon these insects.

The nest was very much damaged by carriage, and Mr. Stone took it entirely to pieces, placing one or two small combs inside a square wooden box with a glass front, and supporting them by a wire which passed through the combs to the roof of the box. He then fixed the box in a window, so as to allow the insects free ingress and egress through a hole in the back.

About three hundred of the workers were then collected,

placed in the box, and well supplied with sugar and beer. They immediately began to work, and their first object was to cover the combs with paper. They worked with great rapidity, and in two days had formed a flask-shaped nest, having covered both the combs and the wire, beside plastering large sheets of paper over the sides of the box. They did not attempt to build upon the glass front, because it was frequently moved in order to introduce a supply of sugar.

As the wasps were building at such a rate, it was evident that they would shortly fill the whole box with a shapeless mass of paper. Another similar box was therefore prepared, and the wasps ejected by tapping the box which was already completed. As soon as they were all out, the second box was substituted for the first, and the wasps crowded eagerly into it and again began their labours. In this box they were allowed to remain for a week, and built another nest. The wasps were now transferred to a third box, in which they laboured for four days, and produced a nest somewhat similar to the others, but not quite so symmetrical.

At this time Mr. Stone fitted up another box with two rows of wire pillars, eight in number, placed with tolerable regularity about two inches apart, and having a piece of comb at the base and summit of each. In this box the wasps remained for fifteen days, and in that time had covered all the wires and most of the combs, and had nearly filled the box with paper.

In order that a more symmetrical structure might be produced, a fifth box was fitted up with wires arranged in a different manner. Four wires were placed across the box, rather in advance of the middle, and two others in front of them. To all these wires a piece of comb was fixed at the base and summit, but between the two central pillars a short wire was placed, having a piece of comb at its summit only. The wasps were transferred to this box, and in the short space of five days they covered all the combs and wires, and produced an extraordinary structure, looking like a paper imitation of a stalactitic cavern. The insects were ejected from this nest before they had finished their work, and in consequence, a portion of the comb on the small central pillar is still left uncovered.

As this box had been so successful, another was prepared on the same principle, and the wasps were permitted to reside in it for the same number of days, in which time they produced an equally beautiful but rather more massive nest. In hopes that the wasps might make a still more splendid nest, a much larger box was fitted up, and the insects transferred to it. As by this time the autumn was closing in, and the weather became cold, the wasps could do but little work, and in a short time they died.

Thus, in the wonderfully short space of thirty-eight days, six elaborate and beautiful nests had been made by a single brood of wasps, and it is probable that if the original nest had been taken at an earlier period of the year, they would have made a still larger number.

IN Mr. Bates's valuable work on the natural history of the Amazons, there is an interesting account of the proceedings of certain ants belonging to the genus *Eciton*, and which are popularly classed together under the name of FORAGING ANTS. These insects have often been confounded with the Saüba or parasol ant, although they belong to different groups and have different habits. The native name for them is Tauóca. There are many species belonging to this genus, and I shall therefore restrict myself to those which seem to have the most interesting habits, giving at the same time a general sketch of their character. I regret that, as in so many other cases, the lack of popular names forces me to employ the scientific titles by which the insects are known to naturalists.

Although in the Ecitons there are the three classes of males, females, and neuters, these neuters are not divided into two distinct sets as in the Termites, but are found in regular gradations of size. The real Foraging Ant is *Eciton drepanophora*, and it is this insect which is so annoying and yet so useful to house-holders. The ants sally forth in vast columns, at least a hundred yards in length, though not of very great width.

On the outside of the column are the officers, which are continually running backwards and forwards, as if to see that their own portions of the column are proceeding rightly. The proportion of officers to workers is about five per cent., or one officer to twenty workers, and they are extremely conspicuous on the march, their great white heads nodding up and down as they run along.

One of the large workers is now before me, and a most formidable insect it looks. Its head is round, smooth, and very large, and is armed with a pair of enormous forceps, curved almost as sharply as the horns of the chamois, and very sharp at the points. Their length is so great, that if straightened and placed end to end, they would be longer than the head and body together. They are beset with minute hairs, which, when viewed under the microscope, are seen to be stiff bristles, arranged in regular rings round the mandibles. The thorax and abdomen are but slender, and the limbs are long, giving evidence of great activity. In the dried specimen, the colour of the insect is yellowish-brown, becoming paler on the head, but when the creature is alive, the head is nearly white. The eyes are very minute, looking like little round dots on the side of the head, and being so extremely small, that they can scarcely be perceived without the aid of a magnifying glass. The half-inch power of the microscope shows that they are oval and convex, but as they are set in little pits or depressions, they do not project beyond the head. The hexagonal compound lenses, which are generally found in insects, are not visible, and the eye bears a great resemblance to that of the spider.

The difference in dimensions of the workers is very remarkable. The specimen which I have just described, measures a little under half an inch in length, exclusive of the limbs, while another specimen is barely half that length, and in general appearance much resembles the familiar ant, or emmet of our gardens.

The presence of these insects may be always known by the numbers of pittas, or ant-thrushes, which feed much upon them, and which are sure to accompany a column of Foraging Ants on the march.

As soon as the experienced inhabitants of tropical America see the ant-thrushes, they rejoice in the coming deliverance, and welcome the approaching army. The fact is, that in those countries insect life swarms as luxuriously as the vegetation, and there are many insects which, however useful in their own place, are apt to get into houses, and there multiply to such an extent, that they become a real plague, and nearly drive the inhabitants out of their own homes. They are bad enough by day, but at night they issue from the nooks and crevices where they lay concealed, and make their presence too painfully known.

There are insects that bite, and insects that suck, and insects that scratch, and insects that sting, and many are remarkable for giving out a most horrible odour. Some of them are cased in armour as hard as crab-shells, and will endure almost any amount of violence, while some are as round, as plump, as thin-skinned, and as juicy as over-ripe gooseberries, and collapse almost with a touch. There are great flying insects which always make for the light, and unless it is defended by glass, will either put it out, or will singe their wings and spin about on the table in a manner that is by no means agreeable. The smaller insects get into the inkstand and fill it with their tiny carcases, while others run over the paper and smear every letter as it is made. There are great centipedes, which are legitimate cause of dread, being armed with poison fangs scarcely less venomous than those of the viper. There are always plenty of scorpions; while the chief army is composed of cockroaches, of dimensions, appetite, and odour such as we can hardly conceive in this favoured land. As to the lizards, snakes, and other reptiles, they are so common as almost to escape attention.

For a time these usurpers reign supreme. Now and then a few dozen are destroyed in a raid, or a person of sanguine temperament amuses his leisure hours, and improves his marksmanship, by picking off the more prominent intruders with a saloon pistol; but the vacancies are soon filled up, and no permanent benefit is obtained. But when the Foraging Ants make their appearance, the case is altered, for there is nothing that withstands their assault. As soon as the pittas are seen approaching, the inhabitants throw open every box and drawer in the house, so as to allow the ants access into every crevice, and then retire from the premises.

Presently the vanguard of the column approaches, a few scouts precede the general body, and seem to inspect the premises, and ascertain whether they are worth a search. The long column then pours in, and is soon dispersed over the house. The scene that then ensues is described as most singular. The ants penetrate into the corners, peer into each crevice, and speedily haul out any unfortunate creature that is lurking therein. Great cockroaches are dragged unwillingly away, being pulled in front by four or five ants, and pushed from behind by as many more. The rats and mice speedily succumb to the onslaught of their myriad foes, the snakes and lizards fare no better, and even the formidable weapons of the scorpion and centipede are overcome by their pertinacious foes.

In a wonderfully short time, the Foraging Ants have completed their work, the scene of turmoil gradually ceases, the scattered parties again form into line, and the procession moves out of the house, carrying its spoils in triumph. The raid is most complete, and when the inhabitants return to the house, they find every intruder gone, and to their great comfort are enabled to move about without treading on some unpleasant creature, and to put on their shoes without previously knocking them against the floor for the purpose of shaking out the scorpions and similar visitors.

Every one who is accustomed to the country takes particular care not to cross one of their columns. The Foraging Ants are tetchy creatures, and not having the least notion of fear, are terrible enemies even to human beings. If a man should happen to cross a column, the ants immediately dash at him, running up his legs, biting fiercely with their powerful jaws, and injecting poison into the wound. The only plan of action in such a case, is, to run away at top speed until the main body are too far off to renew the attack, and then to destroy the ants that are already in action. This is no easy task, for the fierce little insects drive their hooked mandibles so deeply into the flesh that they are generally removed piecemeal, the head retaining its hold after the body has been pulled away, and the mandibles clasped so tightly that they must be pinched from the head and detached separately.

There seems to be scarcely a creature which these insects will not attack, and they will even go out of their way to fall upon the nests of the large and formidable wasps of that country. For the thousand stings the ants care not a jot, but tear away the substance of their nest with their powerful jaws, penetrate into the interior, break down the cells, and drag out the helpless young. Should they meet an adult wasp, they fall upon it, and cut it to pieces in a moment.

I HAVE intentionally reserved the last place among the builders for an insect which is certainly the most wonderful of them all; not only raising an edifice, but clearing a space around, and preparing it for a garden. This insect is called by Dr. Lincecum, the discoverer of its habits, the AGRICULTURAL ANT, and its scientific name is *Atta malefaciens*. As the reader will perceive, it is allied to the parasol ant, which has been already described.

This remarkable insect is a native of Texas, and until a few years ago, its singular habits were unknown. Dr. Lincecum, however, wrote a long and detailed account to Mr. Darwin, who made an abstract of it, and read the paper before the Linnean Society, April 18th, 1861. This abstract may be found in the Journal of that Society, and is as follows:—

'The species which I have named "Agricultural" is a large, brownish ant. It dwells in what may be termed paved cities, and like a thrifty, diligent, provident farmer, makes suitable and timely arrangements for the changing seasons. It is, in short, endowed with skill, ingenuity, and untiring patience, sufficient to enable it successfully to contend with the varying exigencies which it may have to encounter in the life-conflict.

'When it has selected a situation for its habitation, if on ordinary dry ground, it bores a hole, around which it raises the surface three and sometimes six inches, forming a low circular mound, having a very gentle inclination from the centre to the outer border, which on an average is three or four feet from the entrance. But if the location is chosen on low, flat, wet land, liable to inundation, though the ground may be perfectly dry at the time the ant sets to work, it nevertheless elevates the mound, in the form of a pretty sharp cone, to the height of fifteen to twenty inches or more, and makes the entrance near the summit. Around the mound, in either case, the ant clears the ground of all obstructions, and levels and smooths the surface to the distance of three or four feet from the gate of the city, giving the space the appearance of a handsome pavement, as it really is.

'Within this paved area, not a blade of any green thing is allowed to grow, except a single species of grain-bearing grass. Having planted this crop in a circle around, and two or three feet from the centre of the mound, the insect tends and cultivates it with constant care, cutting away all other grasses and weeds that may spring up amongst it, and all around outside the farm-circle to the extent of one or two feet more. The cultivated grass grows luxuriantly, and produces a heavy crop of small, white, flinty seeds, which under the microscope very closely resemble ordinary rice. When ripe, it is carefully harvested and carried by the workers, chaff and all, into the granary cells, where it is divested of the chaff and packed away. The chaff is taken out and thrown beyond the limits of the paved area.

'During protracted wet weather, it sometimes happens that the provision-stores become damp, and are liable to sprout and spoil. In this case, on the first fine day, the ants bring out the damp and damaged grain, and expose it to the sun till it is dry, when they carry it back and pack away all the sound seeds, leaving those that had sprouted to waste.

'In a peach orchard not far from my house is a considerable elevation, on which is an extensive bed of rock. In the sand-beds overlying portions of this rock are fine cities of the Agricultural Ants, evidently very ancient. My observations on their manners and customs have been limited to the last twelve years, during which time the inclosure surrounding the orchard has prevented the approach of cattle to the ant-farms. The cities which are outside the inclosure, as well as those protected in it, are at the proper season invariably planted with the ant-rice. The crop may accordingly always be seen springing up within the circle about the 1st of November every year. Of late years, however, since the number of farms and cattle has greatly increased, and the latter are eating off the grass much closer than formerly, thus preventing the ripening of the seeds, I notice that the Agricultural Ant is placing its cities along the turn-rows in the fields, walks in gardens, inside about the gates, &c., where they can cultivate their farms without molestation from the cattle.

'There can be no doubt that the particular species of grain-bearing grass mentioned above is intentionally planted. In farmer-like manner the ground upon which it stands is carefully divested of all other grasses and weeds during the time it is growing. When it is ripe, the grain is taken care of, the dry stubble cut away and carried off, the paved area being left unencumbered until the ensuing autumn, when the same "ant-rice" reappears within the same circle, and receives the same agricultural attention as was bestowed upon the previous crop—and so on, year after year, as I *know* to be the case, in all situations where the Ants' settlements are protected from graminivorous animals.'

After receiving this account, Mr. Darwin wrote to Dr. Lincecum, asking him whether he thought that the Ants planted seed for the next year's crop, and received the following answer: 'I have not the slightest doubt of it. And, my conclusions have not been arrived at from hasty or careless observation, nor from seeing the Ants do something that looked a little like it, and then guessing the results. I have at all times watched the same ant-cities during the last twelve years, and I know that what I stated in my former letter is true. I visited the same cities yesterday, and found the crop of ant-rice growing finely, and exhibiting also the signs of high cultivation, and not a blade of any other kind of grass or seed was to be seen within twelve inches of the circular row of ant-rice.'

CHAPTER XIX.

SUB-AQUATIC NESTS.

VERTEBRATES.

Fishes as architects—The STICKLEBACKS and their general habits—The FRESH-WATER STICKLEBACKS—A jealous proprietor—Punishment of trespassers—Form and materials of the nest—Use of the nest—Cannibalistic propensities—The FIFTEEN-SPINED STICKLEBACK, and its form—Its curious nest—Mr. Couch's description of a nest in a rope's end.

As a rule, FISHES display but little architectural genius, their anatomical construction debarring them from raising any but the simplest edifice. A fish has but one tool, its mouth, and even this instrument is of very limited capacity. Still, although the nest which a fish can make is necessarily of a slight and rude character, there are some members of that class which construct homes which deserve the name.

The best instances of architecture among the Fishes are those which are produced by the STICKLEBACKS (*Gasterosteus*), those well-known little beings whose spiny bodies, brilliant colours, and dashing courage make them such favourites with all who study nature. There are several species of British Sticklebacks, but as the fresh-water species all make their nests in a very similar manner, there will be no need of describing each species separately.

These fishes make their nests of the delicate vegetation that is found in fresh water, and will carry materials from some little distance in order to complete the home. They do not, however, range to any great extent, because they would intrude upon the preserve of some other fish, and be ruthlessly driven away.

When the male Stickleback has fixed upon a spot for his

nest, he seems to consider a certain area around as his own especial property, and will not suffer any other fish to intrude within its limits. His boldness is astonishing, for he will dash at a fish of ten times his size. and, by dint of his fierce onset and his bristling spears, drive the enemy away. Even if a stick be placed within the sacred circle, he will dart at it, repeating the assault as often as the stick may trespass upon his domains. Within this limit, therefore, he must seek materials for his nest, as he can hardly move for six inches beyond it without intruding upon the grounds of another fish. This right of possession only seems to extend along the banks and a few inches outwards, the centre of the stream or ditch being common property. Along the bank, however, where the vegetation is most luxuriant, there is scarcely a foot of space that is not occupied by some Stickleback, and jealously guarded by him.

Although the nests of the Stickleback are plentiful enough, they are not so familiar to the public as might be expected, principally because they are very inconspicuous, and few of the uninitiated would know what they were, even if they were pointed out. Being of such very delicate materials, and but loosely hung together, they will not retain their form when they are removed from the water, but fall together in an undistinguishable mass, like a coil of tangled thread that had been soaked in water for a few weeks.

The materials of which the nest is made are extremely variable, but they are always constructed so as to harmonise with the surrounding objects, and thus to escape ordinary observation. Sometimes it is made of bits of grass which have been blown into the river, sometimes of straws, and sometimes of growing plants. The object of the nest is evident enough, when the habits of the Stickleback are considered. As is the case with many other fish, there are no more determined destroyers of Stickleback eggs than the Sticklebacks themselves, and the nests are evidently constructed for the purpose of affording a resting-place for the eggs until they are hatched. If a few of these nests be removed from the water in a net, and the eggs thrown into the stream, the Sticklebacks rush at them from all sides, and fight for them like boys scrambling for halfpence. The eggs are very small, barely the size of dust-shot, and are yellow when first placed in the nest, but deepen in colour as they approach maturity.

THERE is a well-known marine species of this group, called the FIFTEEN-SPINED STICKLEBACK (*Gasterosteus spinachia*), a long-bodied, long-snouted fish, with a slightly projecting lower jaw, and a row of fifteen short and sharp spines along the back. This creature makes its nest of the smaller algæ, such as the corallines, and the delicate green and purple seaweeds which fringe our coasts.

Sometimes, indeed, it becomes rather eccentric in its architecture, and builds in very curious situations. Mr. Couch, the well-known ichthyologist, mentions a case where a pair of Sticklebacks had made their nest 'in the loose end of a rope, from which the separated strands hung out about a yard from the surface, over a depth of four or five fathoms, and to which the materials could only have been brought, of course, in the mouth of the fish, from the distance of about thirty feet. They were formed of the usual aggregation of the finer sorts of green and red seaweed, but they were so matted together in the hollow formed by the untwisted strands of the rope, that the mass constituted an oblong ball of nearly the size of the fist, in which had been deposited the scattered assemblage of spawn, and which was bound into shape with a thread of animal substance, which was passed through and through in various directions, while the rope itself formed an outside covering to the whole.'

CHAPTER XX.

SUB-AQUATIC NESTS.

INVERTEBRATES.

A Pool and its wonders—The WATER SPIDER—Its sub-aquatic nest—Conveyance of air to the nest—The diving-bell anticipated—Character of the air in the nest—Mr. Bell's experiment upon the Spider—Life of the Water Spider—The HYDRACHNA—The CADDIS FLIES and their characteristics—Sub-aquatic homes of the Larva—Singular varieties of form and material—Life of a Caddis—Description of nests in my own collection—Fixed cases, and modification of Larva—Singular materials for nest-building—Different species of SABELLA—The SILKWORM AMPHITRITE—The TEREBELLÆ and their submarine houses—The CADDIS SHRIMP—Remarkable analogy.

WHEN I was a very little boy, I was accustomed to spend much time on the banks of the Cherwell, and used to amuse myself by watching the various inhabitants of the water. Animal life is very abundant in that pleasant little river, and there was one favourite nook where a branch of a weeping-willow projected horizontally, and afforded a seat over the dark deep pool, one side of which was abrupt and the other sloping.

Here the merry gyrini ran their ceaseless rounds, and the water-boatmen rowed themselves in fitful jerks, or lay resting in a contemplative manner on their oars. Now and then an unlucky insect would fall from the tree into the water, and then uprose from the dark depth a pair of dull eyes and a gaping mouth, and then, with a glitter as of polished silver, the dace would disappear with its prey. In the shelving part of the pool the caddis-worms moved slowly along, while the great dyticus beetle would rise at intervals to the surface, jerk the end of his tail into the air, and then dive below to the muddy bottom. This spot was much favoured by the nursemaid, for she had no trouble in watching me, as long as I could sit on the branch and look into the water. True, I might have fallen into the river, but I never did ; and even had that accident occurred, it would have wrought no harm, except wet clothes, for I could swim nearly as well as the water-insects themselves.

Close under the bank lived some creatures which always interested me greatly. Spiders they certainly were, but they appeared to have the habits of the water-beetle—coming slowly to the surface of the water, giving a kind of flirt in the air, and then disappearing into the depths, looking like balls of shining silver as they sank down. I had been familiar with these creatures for years before I met with them in some book, and learned that they were known under the name of WATER SPIDER (*Argyronetra aquatica*).

This Spider is a most curious and interesting creature, because it affords an example of an animal which breathes atmospheric air constructing a home beneath the water, and filling it with the air needful for respiration.

The sub-aquatic cell of the Water Spider may be found in many rivers and ditches, where the water does not run very swiftly. It is made of silk, as is the case with all spiders' nests, and is generally egg-shaped, having an opening below. This cell is filled with air ; and if the Spider be kept in a glass vessel, it may be seen reposing in the cell, with its head downwards, after the manner of its tribe. The precise analogy between this nest and the diving-bell of the present day is too obvious to need a detailed account. How the air is introduced into the cell is a problem that was for some time unsolved. The reader is probably aware that the bubbles of air which are to be seen on sub-aquatic plants are almost entirely composed of oxygen gas, which is exuded from the plant, and which is so important an agent in purifying the water. Some zoologists thought that the air which is found in the cell of the Water Spider was nothing but oxygen that had been exuded from the plant upon which the nest was fixed, and that it had been intercepted in its passage to the surface. In order to set the question at rest, Mr. Bell, the well-known naturalist, instituted a series of experiments upon the Spider, and communicated the results to the

Linnean Society. The experiments were made in 1856, and Mr. Bell's remarks are as follow :—

'No. 1. Placed in an upright cylindrical vessel of water, in which was a rootless plant of *Stratiotes*, on the afternoon of

WATER SPIDER.

November 14. By the morning it had constructed a very perfect oval cell, filled with air, about the size of an acorn. In this it has remained stationary up to the present time.

'No. 2. *Nov.* 15. In another vessel, also furnished with Stratiotes, I placed six Argyronetræ. The one now referred to began to weave its beautiful web about five o'clock in the afternoon. After much preliminary preparation, it ascended to the surface, and obtained a bubble of air, with which it immediately and quickly descended, and the bubble was disengaged from the body and left in connexion with the web. As the nest was, on one side, in contact with the glass, inclosed in an angle formed by two leaves of the Stratiotes, I could easily observe all its movements. Presently it ascended again and brought down another bubble, which was similarly deposited.

'In this way, no less than fourteen journeys were performed, sometimes two or three very quickly one after another; at other times with a considerable interval between them, during which time the little animal was employed in extending and giving shape to the beautiful transparent bell, getting into it, pushing it out at one place, and amending it at another, and strengthening its attachments to the supports. At length it seemed to be satisfied with its dimensions, when it crept into it and settled itself to rest with the head downwards. The cell was now the size and nearly the form of half an acorn cut transversely, the smaller and rounded part being uppermost.

'No. 3. The only difference between the movements of this and the former was, that it was rather quicker in forming its cell. In neither vessel was there a single bubble of oxygen evolved by the plant.

'The manner in which the animal possesses itself of the bubble of air is very curious, and, as far as I know, has never been exactly described. It ascends to the surface slowly, assisted by a thread attached to the leaf or other support below and to the surface of the water. As soon as it comes near the surface, it turns with the extremity of the abdomen upwards, and exposes a portion of the body to the air for an instant, then with a jerk it snatches, as it were, a bubble of air, which is not only attached to the hairs which cover the abdomen, but is held on by the two hinder legs, which are crossed at an acute angle near their extremity, this crossing of the legs taking place at the instant the bubble is seized. The little creature then descends more rapidly and regains its cell, always by the same route, turns the abdomen within it, and disengages the bubble.'

The Water Spider places her eggs in this cell, spinning a saucer-shaped cocoon, and fixing it against the inner side of the cell and near the top. In this cocoon are about a hundred eggs, of a spherical shape, and very small. The cell is a true home for the spider, which passes its earliest days under the water, and when it is strong enough to construct a sub-aquatic home for itself, brings its prey to the cell before eating it.

The colour of the Water Spider is brown, with a greyish surface caused by the thick growth of hair which covers the body, and with a very slight tinge of red on the cephalothorax. The reader must not confound this creature with another Arachnid that is sometimes called the Water Spider (*Hydrachna cruenta*), and is of a bright scarlet colour, with a peculiar velvety surface.

THERE is an order of insects which is especially dear to anglers; not so much to fly-fishers, as to those who like to sit and look at a float for several consecutive hours. This order is scientifically termed TRICHOPTERA, or Hair-winged insects, and the various species of which it is composed are classed together under the familiar title of CADDIS FLIES.

These insects may always be known by the peculiar leathery aspect of the body, and by the coating of hair with which the wings are covered, the long hairs being spread over the whole surface, and standing boldly out like a fringe round the edge. They all have long and slender antennæ, and in some genera, such as Mystacida, these organs are nearly three times as long as the head and body.

We will now trace the life of the Caddis Fly from the egg to the perfect insect.

In the breeding season, the female may be observed to carry about with her a double bundle of little greenish eggs, probably in order to expose them for a certain time to the warm sunbeams before they are immersed in the water. This curious bundle is a long oval in shape, and is bent sharply in the middle, its extremities being attached to the abdomen of the insect. When her instinct tells her of the proper time, she proceeds to the water, and attaches the eggs to the leaf of some aquatic plant, often crawling down the stem for several inches. The Caddis Fly is quite at home on the water, and, unlike the dragon flies, which are quite helpless when immersed, can run on the surface with considerable speed, and on occasion can swim below the surface with scarcely less rapidity.

They may often be observed in the act of running on the water, and while they are thus employed, they often fall victims to some hungry fish, which is attracted by the circling ripples occasioned by the movement of the limbs. Fly-fishers, who are acquainted with the habits of fishes and insects, take advantage of their knowledge, and by causing their imitation Caddis Fly to ripple over the surface, or even to sink beneath it, like the veritable insect, delude the unsuspecting fish into swallowing a hook instead of a fly.

In process of time the eggs are hatched, and the young larvæ then proceed to construct houses in which they can dwell. These houses are formed of various materials and are of various shapes, and, indeed, not only does each species have its own particular form of house, but there is considerable variety even in the houses of a single species. In the accompanying illustration are shown a number of the nests formed by the Caddis Fly in its larval state, together with the perfect insects. All the figures have been drawn from actual specimens, some of which are in the British Museum, and others in my own collection. The materials of which the nest is made, depend greatly on the locality in which the insect is hatched, and in a rather large series of Caddis nests now before me, there are some very remarkable instances of the manner in which the insect has been obliged to adapt itself to circumstances. The most common style of case is that which is composed of a number of sticks and grass stems laid longitudinally upon each other like the fasces of the Roman consuls. Of these I have specimens of various sizes and shapes, some being barely half an inch long, while others measure four times that length; the sticks being sometimes placed so irregularly, that the home of the architect is not easily seen. The creatures are not at all particular about the straightness of the sticks, but take them of any degrees of curvature, as in one of

CADDIS.

the examples represented in the illustration, where the stick is not only curved, but has a large bud at the end.

Another case is made of the hollow stem of some plant, apparently that of a hemlock, to which are attached a few slips of bark from the plants. Next comes a series of cases in which the Caddis larva has contrived to secure a great number of cylindrical grass stems and arranged them transversely in several sets, making one set cross the other so as to leave a central space, in which the little architect can live. One or two cases are made wholly of bark, apparently the cuticle of the common reed, a plant which is very common in the Cherwell, whence the cases were taken. In all probability these strips of cuticle have been dropped into the river by the water rats while feeding on the reeds.

Several cases are made entirely of leaves, mostly taken from the white-thorn, which grows in great quantities along the banks of the above-mentioned river. Then, there are cases which are equally composed of sticks and leaves, these materials generally occupying opposite ends of the case. There is another series of cases made up of fine grass, apparently the *débris* of hay which had been blown into the water during the summer, and having the materials laid acoss each other like the needles of a stocking-knitter. Most of these cases are balanced by a stone.

Next come a number of cases which are composed of small shells, those of the Planorbis being the most common, and having among them a few specimens of the Limnæa, or pond-snail, and many separate valves and perfect shells of the fresh-water mussel. The Caddis larva is an incorrigible kidnapper, seizing on any shell that may suit its purpose, without troubling itself about the inhabitant. It is quite a common occurrence to find four or five living specimens of the Planorbis and Limnæa affixed to the case of a Caddis larva, and to see the inhabitants adhering to the plants and endeavouring to proceed in one direction while the Caddis is trying to walk in another, thus recalling the well-known episode of the Tartar and his captor. In these cases the cylindrical body is made of sand and small fragments of shells bound together with a waterproof cement, and the shells are attached by their flat sides to the exterior.

There are also several cases which are made entirely of sand cemented together, some being cylindrical and others tapering to a point, like an elephant's tusk. There are also examples of mixed structures, where the Caddis has combined shells with the leaf and twig cases, and in one of these instances, the little architect has bent back the valves of a small mussel, and fastened them back to back on its house. Beside these, there are one or two very eccentric forms, where the Caddis has chosen some objects which are not often seen in such a position. The seed-vessels of the elm are tolerably common, but I have several specimens where the Caddis has taken the operculum of a dead Pond-snail and fastened it to the case; and there is an example where the chrysalis of some moth, apparently belonging to the genus Porthesia, has been blown into the water from a tree overhanging the stream, and seized upon by a Caddis as an unique ornament for its house. These latter examples were found in a stream in Wiltshire, and the tusk-like sand-cases were found in a disused stone quarry in the same county.

In this remarkable sub-aquatic home the Caddis larva lives in tolerable security, for the head and front of the body are clothed in horny mail, and the soft, white abdomen is protected by the case. The food of the Caddis is generally of a vegetable nature, though there are one or two species which live partly, if not entirely, on animal food. When the larva has lived for its full period, and is about to change into the pupal condition, it closes the aperture of its case with a very strong net, having rather large meshes, and lies securely therein until it is about to change into the winged state. It then bites its way through the net with a pair of strong mandibles, comes to the surface of the water, breaks from its pupal envelope, and shortly takes to flight. The larger species crawl up the stems of aquatic plants before leaving the pupal skin, but the smaller merely stand on the cast skin, which float raft-like on the water.

There are one or two species whose cases are not movable, but are fixed to the spot whereon they were made. In order, therefore, to compensate for the immovability of the case, the larva has a much larger range of movement. In the ordinary species, the creature holds itself to the extremity of the case by means of hooks at the end of its body, which can grasp with some force, as anyone knows who has pulled a Caddis larva out of its house. But when the case is fixed, the abdominal claspers of the larva are attached to a pair of long foot-stalks, so that the creature can extend its body to some distance from the entrance of the tube.

WE now come to some animals that build a submarine edifice, somewhat similar in principle to those of the subaquatic Caddis. The first is the well-known TEREBELLA of our coasts, sometimes known by the name of SHELL-BINDER. Sandy shoals are the best spots for the Terebella, and in many places there is scarcely a square yard of sand without its inhabitants. Like the serpula, the Terebella constructs tubes, but, unlike that animal, it makes the tubes of a soft and flexible texture, although the materials which it employs are harder than those which are used by the serpula. The Terebella has the art of making its submarine tubes of sand, which it agglutinates together with such wonderful power, that if Michael Scott's impish familiar had only been acquainted with natural history, he might soon have learned the art of making ropes of seasand, and have turned the tables on his master.

Should any of my readers be desirous of finding the habitation of a Terebella, he may easily do so by repairing to the nearest sandy shore, and looking under every large stone or piece of rock. There he will probably find some loose tufts of sandy threads, which are fixed to the mouth of a flexible tube, made of the same materials. This tube is the habitation of the Terebella, and by means of a crowbar and a chisel, the animal may generally be procured, together with its home. There are, however, plenty of deserted tubes, and I have often been sadly disappointed by finding that, after a long and laborious digging, nothing but the empty tube was to be found.

Supposing, however, that a specimen is obtained in an uninjured state, the observer can easily watch its method of housebuilding, by ejecting it from its tubular home, placing it in a vessel filled with sea-water, and supplying it with a handful of sand. As clearness of the water is an essential part of success, shell-sand is the best material that can be supplied, and it will be safer to wash the sand thoroughly before placing it in the vessel. A large rough stone should also be placed in the vessel, as the animal always likes to lurk behind some sheltering object while it is engaged in the task of house-building.

Like many other creatures, the Terebella is a night-worker, and during the hours of daylight will retire behind the stone, and crouch in the darkest corner, as if to repose itself after the violent struggles and gyrations which it enacts when it is first taken out of the tube. Until noon is passed, the only sign of life will be the slight movement of the many tentacles which surround the upper lip; but, as the sun declines, the tentacles begin to move more rapidly, and as if they had some purpose to fulfil. In the evening, the worm is in full work; and as Professor Rymer Jones has given a clear and graphic description of its proceedings, I cannot do better than transfer his account to these pages. After remarking on the general habits of the creature, and describing the tentacles, he proceeds as follows:—

'They,' *i.e.* the tentacles, 'are now spread out from the orifice of the tube like so many slender cords—each seizes on one or more grains of sand, and drags its burden to the summit of the tube, there to be employed according to the service required. Should any of the tentacula slip, the same organs are again employed to search eagerly for the lost portion of sand, which is again seized and dragged towards its destination.

'Such operations are protracted during several hours, though so gradually as to be apparently of little effect; nevertheless, on resuming inspection next morning, a surprising elongation of the tube will be discovered; or, perhaps, instead of a simple accession to its walls, the orifice will be surrounded by forking threads of sandy particles agglutinated together.

'The architect has now retired to repose; but as evening comes, its activity is renewed, and again at sunrise a further prolongation has augmented the extent of its dwelling.

'At first sight, the numerous tentacula seem only so many long cylindrical, fleshy threads, of infinite flexibility.

'On examining them, however, more attentively, we see that

in exercising their special function, the surface which is applied to the foreign objects becomes flattened into twice or thrice its ordinary diameter; and while conveying the sandy materials to the tube, these are seized and retained in a deep groove, which almost resembles a slit; in fact, the tentaculum becomes a flat, narrow riband, folding longitudinally in different places to hold the particles securely.

'Although these organs, when contracted, are collected into a brush scarcely double the thickness of the animal's body, so enormous is their extensibility, that they can be stretched out to the length of four inches, or half the length of the body, thus sweeping the area of a circle eight inches in diameter.

'A thin internal coating, resembling silk, lines the whole tube, and at the same time serves as a real cement to unite and strengthen its innumerable parts. This silk-like material is derived from a glutinous slime, which exudes from the surface of the body of the Terebella.

'Notwithstanding the unrivalled expertness and expedition with which this Annelidan advances its work, it has never been observed to resume possession of its tube when once forsaken. To obtain the shelter of a new dwelling in place of the old, its labours are invariably recommenced from the foundation.'

'In *Terebella nebulosa*,' writes Dr. Williams, 'the tentacula consist of hollow, flattened tubular filaments, furnished with strong muscular parietes, each tentacle forming a band which may be rolled longitudinally into a cylindrical form, so as to inclose a hollow, cylindrical space, if the two edges of the band meet, or a semi-cylindrical space, if they imperfectly meet. This inimitable mechanism enables each filament to take up and firmly grasp, *at any point of its length*, a molecule of sand, or, if placed in a linear series, *a row* of molecules. But so perfect is the disposition of the muscular fibres at the extreme free end of each filament, that it is gifted with the twofold power of acting on the sucking and the muscular principle. When the tentacle is about to seize an object, the extremity is drawn in, in consequence of the sudden reflux of fluid in its hollow interior; by this movement a cup-shaped cavity is formed, in which the object is securely held by atmospheric pressure; this power is, however, immediately aided by the contraction of the circular muscular fibres. Such, then, are the marvellous instruments by which these peaceful worms construct their habitations, and probably sweep their vicinity for food.'

It is a remarkable fact that the Terebella does not form tubes during the early portions of its life, but swims about freely, like the nereis and other marine annelids. It has a head, eyes, feet, and antennæ, and roams about at will; whereas, in its perfect state, it has neither head, nor eyes, nor antennæ, nor true feet, the last-mentioned organs being modified into the tufts of hooks, and bristles, by means of which it moves up and down its tube. The reader may perhaps remember that the barnacles and many other stationary marine animals are free during their preliminary epochs, and only become fixed when they attain the perfect form. To our minds, the former seems the more perfect, and certainly the more agreeable state of existence; but we cannot measure the feelings of such an animal by our own, and may be sure that the creature enjoys existence as much while shut up in a tube, as when roaming the ocean at liberty.

ANOTHER species, *Terebella figulus*, sometimes called the POTTER, prefers mud as the material for its dwelling, and contrives to make the dark sea-mud so adhesive that it is capable of being formed into a tube.

As may be easily imagined, this tube is extremely fragile, and cannot be removed entire from the water without the exercise of much care, its own weight being mostly sufficient to tear it asunder. The walls of the tube are tolerably thick, and the tube itself is of some size, measuring nearly half an inch across, and is always found to be protected by the earth upon which it is placed. It is a rather curious fact that the tentacles of this species are of extraordinary length, extending for some eight or nine inches beyond the entrance of the tube, the animal itself measuring little more than four inches in length.

THE last species of Terebella that will be mentioned, is a very small and very remarkable species. It has been appropriately termed the WEAVER TEREBELLA (*Terebella textrix*), from the curious submarine home which it makes.

Not content with using the glutinous secretion as a means for binding together the muddy particles of which the tube is made, it spins a kind of web, bearing some resemblance to that of the spider, and being quite a complicated piece of work. This web is composed of many threads, which are very strong, but are also very fine, and in consequence are almost invisible when in the water, as their substance is quite translucent, like the threads of isinglass. The threads encircle the body, and as the web is only made in the month of May, when the eggs are deposited, it is in all probability employed more for the sake of guarding the eggs than protecting the body.

The tube of the Weaver Terebella is very small, not sufficing to cover more than half the body. The worm seems to be more independent of its tube than is usually the case, frequently vacating and returning to it, and sometimes making two or three tubes near each other, and living in any of them which it may happen to prefer at the time.

WE now come to a group of tube-building annelids which are called Sabellæ, because they live in the sand, and in most cases form their tubes of that material. The general appearance of the tube is extremely variable. In some cases it bears so great a resemblance to the dwelling of the serpula, that a practised eye is needed to discover the distinction.

One very conspicuous species is the TRUMPET SABELLA (*Sabella tubularia*), which is generally found attached to stones or shells. The material of which it is made, is that hard, calcareous matter which is employed by the serpula, and at first the two tubes seem to be exactly alike. A more detailed examination will, however, show that it is not twisted like that of the serpula, but is nearly straight, looking very much like the military trumpet, or 'tuba,' of the ancient Romans. In some cases this tube attains considerable length, measuring eight or nine inches from tip to mouth. It is a solitary animal, and as far as is yet known, is never found grouped in masses, like many allied species.

The gill-fan of this species is exceedingly beautiful, being white, dotted profusely with scarlet, and expanding into a graceful feathery coronet. Although the resemblance to the serpula is very close, the animal may easily be distinguished by the absence of the beautiful operculum or stopper, which forms so conspicuous a feature in the serpula.

PERHAPS the most plentiful species of this genus is the common SABELLA (*Sabella alveolaria*), which may be found in countless myriads on many of our coasts. On several sandy shores, especially those of the southern coast, the wanderer by the sea may perceive masses of hard, agglutinated sand, pierced with innumerable holes. These masses are of great size, and in some places are strong enough to bear the pressure of a foot, though in others a slight push with the hand is sufficient to detach a portion.

If this perforated sand be closely examined, it will be seen to consist of a vast number of tubes, which are fixed together, and are further consolidated by sand which has washed over them, and lodged between them. When the water covers the sand mass, a delicate feathery tuft is seen to protrude from each hole, so that the general aspect is full of beauty. These tufts are the tentacles of the Sabella, and when examined with a microscope of moderate power, each tentacle is seen to be composed of a central shaft, with projecting teeth or fringes on both sides. There are about eighty of these tentacles, and as they are extremely flexible and always in motion, their appearance is peculiarly elegant.

Nothing is easier than to examine the structure of this Sabella, though the task of isolating a single tube is not an easy one. A penknife will soon break up the tube, and a pair of forceps will readily pull out the inhabitant, in spite of the arrray of bristles and hooks wherewith it clings to its habitation. It is but a little creature in point of length, but in point of width it nearly fills the diameter of the tube. The extremity of the body, however, is very small and slender, and is doubled back upon itself, with its tip pointing to the mouth of the tube.

The structure of the tube is extremely variable. Some individuals seem to give all their endeavours towards making their dwelling as long and strong as possible, while others are content with a tube which is barely long enough to shelter the whole body. They work with great rapidity, and when confined in an aquarium, will build their sandy homes nearly as well as if they were at liberty in the sea. Many interesting experiments have been made upon their modes of working, and by a judicious supply of different substances, they may be forced to build tubes of various colours and forms.

THERE is another group of tube-making marine annelids which are remarkable for the transparency of their newly constructed dwellings. Of these a very singular example is found in the SILKWORM AMPHITRITE (*Amphitrite bombyx*).

The reader will remember that one, at least, of the Terebellæ can make a structure which is as transparent as isinglass, and will not, therefore, be surprised to find that another annelid possesses similar powers. The tube of the Silkworm Amphitrite is longer than the body, and is made entirely of the gelatinous secretion which in most of the species is used as a cement for fastening together the sand, shells, mud, and other materials of which the tube is formed. In this creature, however, the secretion is so plentiful, that it forms the whole of the tube.

Nor does it content itself with a single tube, but forms several, one after the other. When first made, the tube is so beautifully transparent, that the body of the inhabitant can be seen almost as plainly as through glass; but in process of time, it becomes incrusted with mud and sand, and almost looks as if it were made of very dirty leather. The average length of an adult specimen is three inches, and its beautiful gill-fan is decorated with brown and yellow. As is the case with most of the tube-inhabiting worms, it is a very timid creature, jerking itself into the tube on the least alarm, and contracting the orifice after it has retired into seclusion.

Should the reader happen to be an entomologist, he will readily call to mind the tiny cylindrical cases that are made by certain lepidopteran larvæ, belonging to the great family Tineidæ, and which are found so plentifully upon the leaves of oak, hazel, and other trees. If he should happen to be something of an aquarian naturalist, and fond of looking for marine curiosities, he may find attached to submarine plants, certain little cylindrical cases which are wonderfully like those of the moths. They are very small indeed, scarcely thicker than the shaft of an ordinary pin, and measuring scarcely more than the eighth of an inch in length. Their colour is pale brown, their surface is rough, and they are stuck upon the seaweed in great confusion, without the least attempt at arrangement.

These are the habitations of a very small crustacean (*Cerapus tubularis*), popularly called the Caddis Shrimp, because the tube which the creature makes is analogous to that which is formed by the caddis larvæ. The animal which inhabits this case is a curious little being, very like the long-bodied, long-legged, caprellæ, that are so plentiful among seaweeds, and furnished with two pairs of long and stout antennæ, and two pairs of grasping feet. As the tube is too short to contain the entire animal, the long antennæ are always protruded, and occasionally the powerful grasping feet are also thrust out of the opening.

The antennæ are continually flung forward and retracted in a manner that reminds the observer of the movements of the acorn barnacle, each grasp being evidently made for the purpose of arresting any passing substance that may serve for food. This remarkable little crustacean is generally found upon the well-known alga which produces the Carrageen, or Irish moss (*Chondrus crispus*). It will not, however, be found upon those plants which can be plucked by hand, but resides in deeper water, so that the best method of procuring it is to go out in a boat, throw the drag overboard, and then examine the algæ which are torn from their attachments.

CHAPTER XXI.

SOCIAL HABITATIONS.

SOCIAL MAMMALIA.

The BEAVER—Its form and aquatic habits—Need for water and means used to procure it—Quadrupedal engineering—The dam of the Beaver—Erroneous ideas of the dam—How the Beaver cuts timber—The Beaver in the Zoological Gardens—Theories respecting the Beaver's dam—How the timber is fastened together—Form of the dam, and mode of its enlargement—Beaver-dams and coral-reefs—The house or lodge of the Beaver—Its locality and structure—Use of a subterranean passage—How Beavers are hunted—Curious Superstition—'Les Paresseux.'

WE now come to the SOCIAL HABITATIONS, and give precedence to those which are constructed by Mammalia.

Of the Social Mammalia, the BEAVER (*Castor fiber*) takes the first rank, and is the best possible type of that group. There are other social animals, such as the various marmots and others; but these creatures live independently of each other, and are only drawn together by the attraction of some favourable locality. The Beavers, on the other hand, are not only social by dwelling near each other, but by joining in a work which is intended for the benefit of the community.

The form of the Beaver is sufficiently marked to indicate that it is a water-loving creature, and that it is a better swimmer than walker. The dense, close, woolly fur, defended by a coating of long hairs, the broad, paddle-like tail, and the well-webbed feet, are characteristics which are at once intelligible. Water, indeed, seems to be an absolute necessity for the Beaver, and it is of the utmost importance to the animal that the stream near which it lives, should not be dry. In order to avert such a misfortune, the Beaver is gifted with an instinct which teaches it how to keep the water always at or about the same mark, or, at all events, to prevent it from sinking below the requisite level.

If any modern engineer were asked how to attain such an object, he would probably point to the nearest water-mill, and say that the problem had there been satisfactorily solved, a dam having been built across the stream so as to raise the water to the requisite height, and to allow the superfluous water to flow away. Now, water is as needful for the Beaver as for the miller, and it is a very curious fact, that long before millers ever invented dams, or before men ever learned to grind corn, the Beaver knew how to make a dam and insure itself a constant supply of water.

That the Beaver does make a dam is a fact that has long been familiar, but how it sets to work is not so well known. Engravings representing the Beavers and their habitations, are common enough, but they are generally untrustworthy, not having been drawn from the natural object, but from the imagination of the artist. In most cases the dam is represented as if it had been made after the fashion of our time and country, a number of stakes having been driven into the bed of the river, and smaller branches entwined among them. The projecting ends of the stakes are neatly squared off, and altogether the work looks exactly as if it had been executed by human hands. One artist seems to have copied from another, so that the error of one man has been widely perpetuated by a series of successors.

Now, in reality, the dam is made in a very different manner, and in order to comprehend the mode of its structure, we must watch the Beaver at work.

When the animal has fixed upon a tree which it believes to be suitable for its purpose, it begins by sitting upright, and with its chisel-like teeth, cutting a bold groove completely round the trunk. It then widens the groove, and always makes it wide in exact proportion to its depth, so that when the tree is nearly cut through, it looks something like the contracted portion of an hour-glass. When this stage has been reached, the Beaver looks anxiously at the tree, and views it on every side, as if desirous of measuring the direction in which it is to fall. Having settled this question, it goes to the opposite side of the tree, and with two or three powerful bites cuts away the wood, so that the tree becomes overbalanced and falls to the ground.

This point having been reached, the animal proceeds to cut up the fallen trunk into lengths, usually a yard or so in length, employing a similar method of severing the wood. In consequence of this mode of gnawing the timber, both ends of the logs are rounded and rather pointed. In the Zoological Gardens may be seen many excellent examples of timber which has been cut by the Beaver. The logs and stumps which project a foot or so from the ground are so neatly pointed that very few visitors notice them, thinking them to be cut by the hand of man.

The next part of the task is, to make these logs into a dam. Now, whereas some persons have endeavoured to make the Beaver a more ingenious animal than it really is, and have accredited it with powers which only belong to mankind, others have gone to the other extreme, and have denied the existence of a regularly built dam, saying that it is entirely accidental, and caused by the logs that are washed down by the stream, after the Beavers have nibbled off all the bark.

That this position is untenable is evident from the acknowledged fact that the dam is by no means placed at random in the stream, just where a few logs may have happened to lodge, but is set exactly where it is wanted, and is made so as to suit the force of the current. In those places where the stream runs slowly, the dam is carried straight across the river, but in those where the water has much power, the barrier is made in a convex shape, so as to resist the force of the rushing water. The power of the stream can, therefore, always be inferred from the shape of the dam which the Beavers have built across it.

Some of these dams are of very great size, measuring two or three hundred yards in length, and ten or twelve feet in thickness, and their form exactly corresponds with the force of the stream, being straight in some parts, and more or less convex in others.

The dam is formed, not by forcing the ends of the logs into the bed of the river, but by laying them horizontally, and covering them with stones and earth until they can resist the force of the water. Vast numbers of logs are thus laid, and as fast as the water rises, fresh materials are added, being obtained mostly from the trunks and branches of trees which have been stripped of their bark by the Beavers.

The reader will remember that many persons have thought that the dam of the Beaver is only an accidental agglomeration of loose logs and branches, without any engineering skill on the part of the animals. There is some truth in this statement, though the assertion is too sweeping. For, after the Beavers have completed their dam, it obstructs the course of the stream so completely that it intercepts all large floating objects, and every log or branch that may happen to be thrown into the river is arrested by the dam, and aids in increasing its dimensions.

Mud and earth are also continually added by the Beavers, so that in process of time the dam becomes as firm as the land through which the river passes, and is covered with fertile alluvium. Seeds soon make their way to the congenial soil, and in a dam of long standing, forest trees have been known to grow, their roots adding to the general stability by binding together the materials. It is well known that the fertile islands formed on coral reefs are stocked in a similar manner. Originally, the dam is seldom more than a yard in width where it overtops the water, but these unintentional additions cause a continual increase.

The bark with which the logs were originally covered, is not all eaten by the animals, but stripped away, and the greater part hidden under water, to serve for food in the winter time. A further winter provision is also made by taking the smaller branches, diving with them to the foundations of the dam, and carefully fastening them among the logs. When the Beavers are hungry, they dive to their hidden stores, pull out a few branches, carry them on land, nibble away the bark, and drop the stripped logs on the water, where they are soon absorbed by the dam.

We have now seen how the Beavers keep the water to the required level, and we must next see how they make use of it.

The Beaver is essentially an aquatic mammal, never walking when it can swim, and seldom appearing quite at its ease upon dry land. It therefore makes its houses close to the water, and communicating with it by means of subterranean passages, one entrance of which passes into the house or 'lodge,' as it is technically named, and the other into the water, so far below the surface that it cannot be closed by ice. It is, therefore, always possible for the Beaver to gain access to the provision stores, and to return to its house, without being seen from the land.

The lodges are nearly circular in form, and much resemble the well-known snow houses of the Esquimaux, being domed, and about half as high as they are wide, the average height being three feet and the diameter six or seven feet. These are the interior dimensions, the exterior measurement being much greater, on account of the great thickness of the walls, which are continually strengthened with mud and branches, so that, during the severe frosts, they are nearly as hard as solid stone. Each lodge will accommodate several inhabitants, whose beds are arranged round the walls.

All these precautions are, however, useless against the practised skill of the trappers. Even in winter time the Beavers are not safe. The hunters strike the ice smartly, and judge by the sound whether they are near an aperture. As soon as they are satisfied, they cut away the ice and stop up the opening, so that if the Beavers should be alarmed, they cannot escape into the water. They then proceed to the shore, and by repeated soundings, trace the course of the Beavers' subterranean passage, which is sometimes eight or ten yards in length, and by watching the various apertures are sure to catch the inhabitants. This is not a favourite task with the hunters, and is never undertaken as long as they can find any other employment, for the work is very severe, the hardships are great, and the price which they obtain for the skins is now very small.

While they are thus engaged, they must be very careful not to spill any blood, as if they do so, the rest of the Beavers take alarm, retreat to the water, and cannot be captured. They also have a curious superstitious notion, which leads them to remove a knee-cap from each Beaver and to throw it into the fire. They would expect ill-luck were they to omit this ceremony, which is wonderfully like the custom of our fishermen of spitting into the mouth of the first fish they catch, and on the first money which they take in the day, 'for luck.'

Generally, the Beavers desert their huts in the summer time, although one or two of the houses may be occupied by a mother and her young offspring. All the old Beavers who have no domestic ties to chain them at home, take to the water, and swim up and down the stream at liberty, until the month of August, when they return to their homes. There are, also, certain individuals called by the trappers 'les paresseux,' or idlers, which do not live in houses, and make no dam, but abide in subterranean tunnels like those of our common water rat, to which they are closely allied. These 'paresseux' are always males, and it sometimes happens that several will inhabit the same tunnel. The trapper is always pleased when he finds the habitation of an idler, as its capture is a comparatively easy task.

CHAPTER XXII.

SOCIAL BIRDS.

The SOCIABLE WEAVER BIRD and its country—Description of the bird—Nest of the Sociable Weaver—How begun and how carried on—Materials of the nest—The tree on which the nest is built, and its uses—Dimensions of the nest and disastrous consequences—A Hottentot and a lion—Supposed object of the Social nest—Average number of inhabitants—Enemies of the Sociable Weaver, the monkey, the snake, and the parrakeet.

WE now come to the SOCIAL BIRDS, one of which is as pre-eminent among the feathered tribes as is the beaver among mammalia. This is the SOCIABLE WEAVER BIRD, sometimes called the SOCIABLE GROSBEAK (*Philetærus socius*).

This species is allied to the Weaver Birds, some of which have already been described, and makes a nest which is no whit inferior to those which have already been mentioned. The Sociable Weaver Bird is a native of Southern Africa, and in some places is very plentiful, its presence depending much upon the trees which clothe the country. It is not a large bird, measuring about five inches in length, and is very inconspicuous, its colour being pale bluff, mottled on the back with deep brown.

The chief interest about the species is concentrated in its nest, which is a wonderful specimen of bird architecture, and attracts the attention of the most unobservant traveller. Few persons expect to see in a tree a nest which is large enough to shelter five or six men; and yet that is often the case with the nest of the Sociable Weaver Bird. Of course so enormous a structure is ot the work of a single pair, but, like the dam of the beaver, is made by the united efforts of the community. How it is built will now be described.

Large as is the domicile, and capable at last of containing a vast number of parents and young, it is originally the work of a single pair, and attains its enormous dimensions by the labours of those birds which choose to associate in common. The first task of this Weaver Bird is to procure a large quantity of the herb which really seems as if made expressly for the purpose.

SOCIABLE WEAVER BIRD.

This is a grass with a very large, very tough, and very wiry blade, which is known to the colonists as Booschmannie grass, probably because it grows plentifully in that part of Southern Africa where the Bushmen, or Bosjesmans live.

They carry this grass to some suitable tree, which is usually a species of acacia, called by the Dutch colonists Kameeldorn (*Acacia giraffa*), because the giraffe, which the Dutch persist in calling a kameel or camel, is fond of grazing on the leaves.

The birds then hang the Booschmannie grass over a suitable branch, and by means of weaving and plaiting it, they form a roof of some little size. Under this roof are placed a quantity of nests, increasing in number with each successive brood. The nests are set closely together, so that at last they look like a mass of grass pierced with numerous holes, and it is really wonderful that the birds should be able to find their way to their own particular homes. To human eyes, the nests are as much alike as the houses in a modern street, before the blinds, the flowers, and other additions have communicated an individuality to each dwelling; but, notwithstanding this similarity, the inmates glide in and out without any hesitation.

Although the same nest-mass is occupied for several successive seasons, the birds refuse to build in the same nests a second time, preferring to make a fresh domicile for each new brood. In consequence of this custom, when the birds have entirely filled the roofs with their nests, they do not desert it, but enlarge the roof, and build a second row of nests, just like the combs of a wasp's or hornet's habitation.

Layer after layer is thus added, until the mass becomes of so enormous a size that travellers have mistaken these nests for the houses of human beings, and been grievously disappointed when they came near enough to detect their real character. There is a story of a Hottentot and a lion, which will give an idea of the dimensions of these nests. A Hottentot, who was engaged in some task, was suddenly surprised by a lion, and instinctively made for the nearest tree, which happened to be a kameel-dorn. Up the tree he sprang, and finding one of the branches occupied by the nest of the Sociable Weaver Bird, he took refuge behind the grassy mass, and was thus concealed from the pursuer.

The lion, in the meantime, arrived at the foot of the tree,

but could not see his intended prey. The unlucky Hottentot, however, peeped over the nest in order to see whether the coast was clear, and was spied by the lion, who made a dash at the tree. The man shrank back behind the nest, but his imprudent movement brought its own punishment.

Knowing that the ascent of the tree was impossible, and at the same time unwilling to leave its prey, the lion sat down at the foot of the tree, and kept watch upon the man. Hour after hour the lion mounted guard over its prisoner, until thirst overpowered hunger, and the animal was forced reluctantly to quit its post and seek for water. The man then scrambled down the tree, and made the best of his way homewards, little the worse for his imprisonment except the fright, and a skin scorched by long exposure to the sun. The artist has introduced this little episode into the illustration, because it enables the reader to judge of the enormous size of the nest.

Season after season the Weaver Birds continue to add their nests, until at last the branch is unable to endure the weight, and comes crashing to the ground. This accident does not often occur during the breeding months, but mostly takes place during the rainy season, the dried grass absorbing so much moisture, that the weight becomes too great for the branch to bear.

The nest group which is shown in the illustration is of medium size, as can be ascertained by its shape. In its early state, the nest-mass is comparatively long and narrow, spreading out by degrees as the number of nests increases, so that at last it is as wide and as shallow as an extended umbrella. The dimensions of some of these structures may be gathered from the fact, that Le Vaillant counted in one unfinished edifice, beside the deserted nests of previous seasons, no less than three hundred and twenty nests, each of which was occupied by a pair of birds engaged in bringing up a brood of young, four or five in number.

The Weaver Birds have but few enemies. First, there are the snakes, which are such determined robbers of nests, swallowing both eggs and young ; and then there are the monkeys, which are capable of sad depredations whenever they can find an opportunity. Monkeys are extremely fond of eggs, and there is scarcely a better bribe to a monkey, ape, or baboon, than a fresh raw egg. The bird which laid it is almost as great a dainty, and a monkey seems to be in the height of enjoyment if a newly-killed bird be put into its paws. It always begins by eating the brain, and then tears the carcase to pieces with great deliberation. A mouse is quite as much appreciated as a bird, provided that it has been recently killed, and that the blood has not congealed.

However, the structure of the nest forms an insurmountable barrier to the snake, and the monkey can only reach a few of the cells which are near the edge. The worst enemies are certain little parrakeets, which are delighted to be able to procure nests without the trouble of building them, and which are apt to take possession of the cells and oust the rightful owners.

CHAPTER XXIII.

SOCIAL INSECTS.

Nests of POLYBIA—Curious method of enlargement—Structure of the nests—How concealed—Various modes of attachment—A curious specimen—The HIVE BEE, and claims to notice—General history of the hive—Form of the cells—The royal cell, its structure and use—Uses of the ordinary cells—Structure of the Bee-cell—Economy of space—How produced—Measurement of angles—A logarithmic table corrected by the bee-cell—The 'lozenge,' a key to the cell—How to form it—Beautiful mathematic proportions of the lozenge—Method of making the cell or a model—Effect of the cell upon honey—The HORNET and its nest—Its favourite localities—Difficulties of taking a hornet's nest—Habits of the insect—Mr. Stone's method of taking the nest—The SMALL ERMINE MOTH—and its ravages—Its large social habitation—General habits of the larva—The GOLD-TAILED MOTH, and its beautiful social nest—Description of a specimen from Wiltshire—Illustration of the theory of heat—The BROWN-TAILED MOTH and its nest.

AFTER the Social Birds come the SOCIAL INSECTS, to which the following chapter is dedicated.

Just as the hymenoptera are chief among the pensiles and the builders, so are they chief among the Social Insects, and the species which may be placed in this group are so numerous, that it will only be possible to make a selection of a few, which seem more interesting than the others.

IN the British Museum there are some very remarkable nests made by hymenopterous insects belonging to the genus *Polybia*, several of which are drawn in the accompanying illustration. As it was desirable to include more than one specimen, the figures are necessarily much reduced in size. Neither the nests nor the insect, however, are of large dimensions, and the former are so sombre in colour as well as small in size, that they would not of themselves attract any attention. Their nests, however, are extremely interesting, as may be seen from the examples which are figured in the illustration.

On the left hand may be seen a nearly spherical nest, which is evidently hollow, and has cells both on the outside and within the cover. These cells are not placed vertically, with their mouths downward, like those of the wasp and hornet, nor

POLYBIA.

horizontally like those of the bee, but are set with their mouths radiating from the centre of the nest. Moreover, there is another curious circumstance connected with the nest. If it were to be opened, it would be seen to be composed of several concentric layers, very much like those ivory puzzle-balls which the Chinese make so beautifully.

The method by which the nest is formed is very simple, though not one that is usually seen among the hymenoptera. The layers of combs are made like hollow spheres, the mouths of the cells being outwards, and as soon as a layer is completed, the insects protect it from the weather by a cover of the same material as is used for the construction of the cells. When they require to make a fresh layer of cells, they do not enlarge the cover, as is the case with the wasp and hornet, but place the new cells upon the surface of the cover, and make a fresh cover as soon as the comb is completed. Thus the nest increases by the addition of concentric layers, composed alternately of comb and cover.

In the nest which is in the British Museum, the insects have commenced several patches of comb on the outside of the cover, and one such patch is shown in the illustration.

On the right of the globular nest is another curious structure, also made by insects of the same genus, and having a kind of similarity in its aspect. This nest, however, is very much longer in proportion to its width, and being fixed throughout its length to a leaf, is not so plainly visible as the last-mentioned specimen. Indeed, when the leaf has withered, as is the case with the object from which the drawing was made, the dull brown of the nest coincides so completely with the colour of the faded leaf, that many persons would overlook it unless their attention were specially drawn towards it.

On the extreme right of the illustration, and in the upper corner, is seen a nest which is also the work of insects belonging to the genus Polybia, and it is pendent from a bough, like the habitation of the Chartergus and other pensile hymenoptera.

In the same collection there are many more specimens of social nests formed by insects belonging to this genus, two cases being quite filled with them. One is attached to the bark of a tree, and resembles it so closely that it seems to be made of the same substance, this similarity of aspect being evidently intended as a preservative against the attacks of birds and other insect-loving creatures, which would break up the nest, and eat the immature and tender grubs. Most of the nests are fixed to leaves, and are different forms, according to the species which made them. They are mostly fixed to the under sides of the leaf, so that the weight causes the leaf to bend and to form a natural roof above them. The shape of the nest seems to depend much on the character of the plant to which it is fixed. Those that are fastened to reeds are long and slender, and generally much narrower than the sword-shaped leaf on which they rest. Others, which are fastened to short and broad leaves, adapt themselves so closely to the shape of the leaf, that, if removed, they would enable anyone to conjecture the form of the leaf upon which they had been fixed.

One such nest is very remarkable. In general form it bears a singular resemblance to the nest of the fairy martin, though its materials are entirely different. The nest is flask-shaped, and its base is fastened to a leaf which it almost covers. The body of the nest is oval, and the entrance, which is small, is placed at the end of a well-marked neck. The shell of the nest is extremely thin, not in the least like the loose, papery structure of an ordinary wasp-nest, nor the pasteboard-like material which defends the nest of the Chartergus. It is rather fragile, and in thickness is almost double that of the paper on which this account is printed.

The name of the species which builds this curious nest is *Polybia sedula*, and the specimen was brought from Brazil.

For the reasons which have been given at the beginning of this chapter, the Hive Bee has been reckoned among the Social Insects.

The Bee has always been one of the most interesting insects to mankind, on account of the direct benefit which it confers upon the human race. There are many other insects which are in reality quite as useful to us, and indeed are indispensable, but which we neglect because we are ignorant of their labours. The Bee, however, furnishes two powerful and tangible arguments in its favour—namely, honey and wax—and is sure, therefore, to enlist our sympathies in its behalf.

Independently, however, of these claims to our notice, if the Bee never made an ounce of honey—if the wax were as useless to us as wasp-comb—if the insect were a mere stinging creature, with a tetchy temper, it would still deserve our admiration, on account of the wonderful manner in which it constructs its social home, and the method by which that home is regulated.

I need not in this place repeat the well-known facts respecting the constitution of the Bees, nor describe the duties of the Queen, Drone, and Worker Bees. Suffice it to say, that the former is the mother as well as the queen of the hive; that the workers are undeveloped females, which are properly called neuters; and that the drones are males, which do no work, and have no stings.

In the Queen Bee, the abdomen is long in proportion to its width, and the wings slightly cross each other when closed; the latter being a very conspicuous badge of sovereignty. The drones are easily distinguished by their generally larger size, their larger eyes, and the wide, blunt, and rounded abdomen.

There are three kinds of cell in a hive; namely, the worker-cell, the drone-cell, and the royal-cell. Of these, the two former are hexagonal, but can easily be distinguished by the greater size of the drone-cell; while the royal-cell is totally unlike the nursery of a subject, whether drone or worker, and is almost always placed on the edge of a comb.

The little grub which is placed in the royal-cell is not fed with the same food which is supplied to the other Bees, but lives upon an entirely different diet, and which is, apparently, of a more stimulating character; and it is now well known, that if a young grub which has been hatched in one of the worker-cells be removed into the royal-cell, and supplied with royal food, it becomes developed into a queen, and, in time, is qualified to rule and populate a hive. This remarkable provision of nature is intended to meet a difficulty, which sometimes occurs, when the reigning queen dies, and there is no royal larva in the cell.

The chief point which distinguishes the comb of the Hive Bee from that of other insects, is the manner in which the cells are arranged in a double series. The combs of the wasp or the hornet are single, and are arranged horizontally, so that their cells are vertical, with the mouths downwards and the bases upwards, the united bases forming a floor on which the nurse wasps can walk while feeding the young inclosed in the row of cells immediately above them.

Such, however, is not the case with the Hive Bee. As every one knows, who has seen a bee-comb, the cells are laid nearly horizontally, and in a double series, just as if a couple of thimbles were laid on the table with the points touching each other and their mouths pointing in opposite directions. Increase the number of thimbles, and there will be a tolerable imitation of a bee-comb.

There is another point which must now be examined. If the bases of the cells were to be rounded like those of the thimbles, it is clear that they would have but little adhesion to each other, and that a large amount of space would be wasted. The simplest plan of obviating these defects is evidently to square off the rounded bases, and to fill up the ends of each cell with a hexagonal flat plat, which is actually done by the wasp. If, however, we look at a piece of bee-comb, we shall find that no such arrangement is employed, but that the bottom of each cell is formed into a kind of three-sided cup. Now, if we break away the walls of the cells, so as only to leave the bases, we shall see that each cup consists of three lozenge-shaped plates of wax, all the lozenges being exactly alike.

These lozenge-shaped plates contain the key to the bee-cell, and their properties will therefore be explained at length. Before doing so, I must acknowledge my thanks to the Rev. Walter Mitchell, Vicar and Hospitaller of St. Bartholomew's Hospital, who has long exercised his well-known methematical powers on this subject, and has kindly supplied me with the outline of the present history.

If a single cell be isolated, it will be seen that the sides rise from the outer edges of the three lozenges above-mentioned, so that there are, of course, six sides, the transverse section of which gives a perfect hexagon. Many years ago Maraldi, being struck with the fact that the lozenge-shaped plates always had the same angles, took the trouble to measure them, and found that in each lozenge, the large angles measured 109° 28′, and the smaller, 70° 32′, the two together making 180°, the equivalent of two right angles. He also noted the fact that the apex of the three-sided cup was formed by the union of three of the greater angles. The three united lozenges are seen at fig. 1.

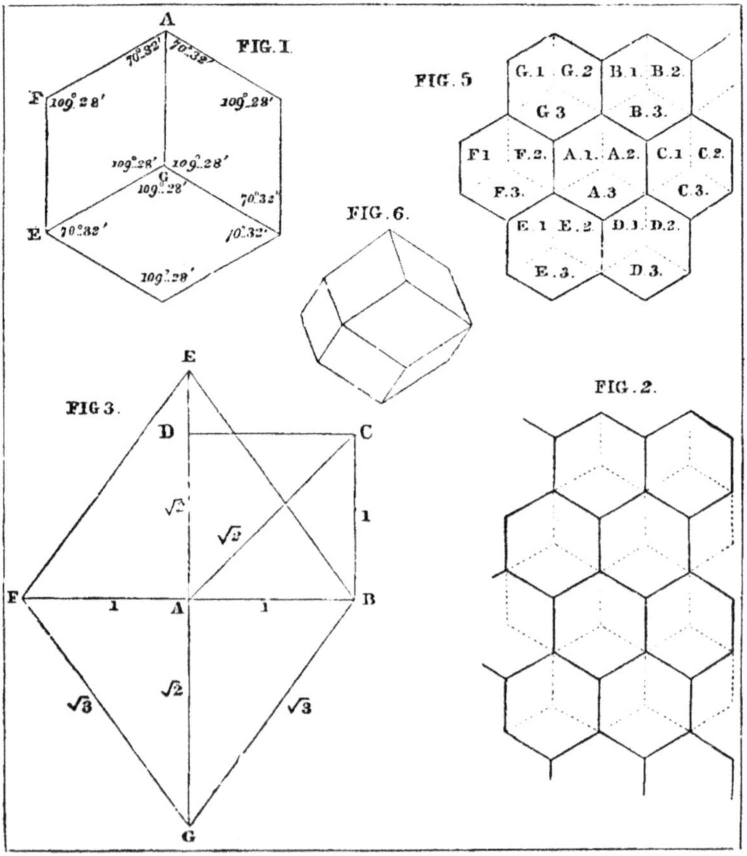

Some time afterwards, Reaumur, thinking that this remarkable uniformity of angle might have some connection with the wonderful economy of space which is observable in the bee-comb, hit upon a very ingenious plan. Without mentioning his reasons for the question, he asked Kœnig, the mathematician, to make the following calculation. Given a hexagonal vessel terminated by three lozenge-shaped plates; what are the angles which would give the greatest amount of space with the least amount of material?

Kœnig made his calculations, and found that the angles were 109° 26′ and 70° 34′, almost precisely agreeing with the measurements of Maraldi. The reader is requested to remember these angles. Reaumur, on receiving the answer, concluded that the Bee had very nearly solved the difficult mathematical problem, the difference between the measurement and the calculation being so small as to be practically negatived in the actual construction of so small an object as the bee-cell.

Mathematicians were naturally delighted with the result of the investigation, for it showed how beautifully practical science could be aided by theoretical knowledge, and the construction of the bee-cell became a famous problem in the economy of nature. In comparison with the honey which the cell is intended to contain, the wax is a rare and costly substance, secreted in very small quantities, and requiring much time for its production; it is therefore essential that the quantity of wax employed in making the comb should be as little, and that of the honey contained in it as great, as possible.

For a long time these statements remained uncontroverted. Anyone with the proper instruments could measure the angles for himself, and the calculations of a mathematician like Kœnig would hardly be questioned. However, Maclaurin, the well-known Scotch mathematician, was not satisfied. The two results very nearly tallied with each other, but not quite, and he felt that in a mathematical question precision was a necessity. So he tried the whole question himself, and found Maraldi's measurements correct, namely, 109° 28′, and 70° 32′.

He then set to work at the problem which was worked out by Kœnig, and found that the true theoretical angles were 109° 28′, and 70° 32′, precisely corresponding with the actual measurement of the bee-cell.

Another question now arose. How did this discrepancy occur? How could so excellent a mathematician as Kœnig make so grave a mistake? On investigation, it was found that no blame attached to Kœnig, but that the error lay in the book of logarithms which he used. Thus, a mistake in a mathematical work was accidentally discovered by measuring the angles of a bee-cell—*a mistake sufficiently great to have caused the loss of a ship whose captain happened to use a copy of the same logarithmic tables for calculating his longitude.*

Now, let us see how this beautiful lozenge is made. There is not the least difficulty in drawing it. Make any square, ABCD (fig. 3) and draw the diagonal AC.

Produce BA towards F and AD, both ways to any distance.

Make AE and AG equal to AC, and make AF equal to AB. Join the points EFGB, and you have the required figure.

Now comes a beautiful point. If we take AB as 1, being one side of the square on which the lozenge is founded, AE and AG will be equal to $\sqrt{2}$, and EF, FG, GB, and BE, will be equal to $\sqrt{3}$, as can be seen at a glance by anyone who has advanced as far as the 47th proposition of the first book of Euclid.

We have not yet exhausted the wonders of the bee-comb.

If we take a piece of comb from which all the cells have been removed, and hold it up to the light, we shall see that the cells are not placed opposite each other, but that the three lozenges which form the base of one cell form part of the base of three other cells, as is seen in fig. 2.

It would, of course, be easy to fill many pages with the account of the Hive Bee and its habits; but as this work is restricted to the habitations of animals, we can only look upon the Bee as a maker of social habitations. It will, however, be necessary to mention the material of which the comb is made.

The other hymenoptera obtain their materials from external sources. The hornet and wasp have recourse to trees and branches, and bear home in their mouths the bundles of woody fibres which they have gnawed away. The upholsterer and leaf-cutter Bees are indebted to the petals and leaves of various plants, and various wood-boring insects make their homes of the woody particles which they have nibbled away. The Bee, however, obtains her wax in a very different manner.

If the body of a worker Bee be carefully examined, on the under sides of the abdomen will be seen six little flaps, not unlike pockets, the covers of which can be easily raised with a pin or needle. Under these flaps is secreted the wax, which is produced in tiny scales or plates, and may be seen projecting from the flaps like little semilunar white lines. Plenty of food, quiet, and warmth are necessary for the production of wax, and as it is secreted very slowly, it is so valuable that the greatest economy is needed in its use. It is, indeed, a wonderful substance; soft enough when warm to be kneaded and to be spread like mortar, and hard enough when cold to bear the weight of brood and honey. Moreover, it is of a texture so close that the honey cannot soak through the delicate walls of the cells, as would soon be the case if the comb were made of woody fibre, like that of the hornet or wasp.

Indeed, it is a most remarkable fact that the Bee should be able to produce not only the honey, but the material with which is formed the treasury wherein the honey is stored. Honey itself is again scarcely less remarkable than wax. The Bee goes to certain flowers, inserts its hair-clad proboscis into their recesses, sweeps out the sweet juice, passes the laden proboscis through its jaws, scrapes off the liquid and swallows it. The juice then passes into a little receptacle just within the abdomen called the 'honey-bag,' which is apparently composed of an exceedingly delicate membrane, and seems to discharge no other office than that of a vessel in which the juice can be kept while the Bee is at work.

As soon as the honey-bag is filled, the Bee flies back to the hive and disgorges the juice into one of the cells. But, during that short sojourn in the insect, the juice has undergone a change, and been converted into honey, a substance which is quite unlike that from which it was formed, and which has an odour and flavour peculiarly its own. How this change is wrought is at present unknown, for the little bag in which the transformation is made is composed of a membrane that seems incapable of exerting any influence upon the substance contained within it.

All food that is eaten by the Bee passes through the honey-bag, which is closely analogous to the crop of a bird, and it would seem that the honey ought rather to pass into the stomach than be disgorged at the will of the insect. However, it is well known that many birds feed their young by disgorging food, and the Bee is enabled to perform the same operation by means of a little valve which leads from the honey-bag into the stomach, and is plainly perceptible even with the unassisted eye. Under ordinary circumstances the valve just allows the food to pass gently and gradually into the stomach ; but the violent effort, which is made in ejecting the food, closes the valve, and only allows the honey to flow upwards through the mouth.

The office of the worker and drone cells is two-fold—first, to act as nurseries for the insects while passing through their preliminary stages, and next to serve as repositories for food, whether liquid or solid. The egg of the Queen-Bee is placed nearly at the bottom of the cell, exactly on the angle where the point of the lozenges meet. It is soon hatched into a little white grub, which is assiduously fed by the nurses, and grows with wonderful rapidity. As soon as it has eaten its last larval meal, it spins a silken cover over the cell, and remains there until it has become a perfect insect. It then bites its way out, and after a day or so devoted to hardening and strengthening its limbs, it leaves the hive and joins in the labours of the community.

No sooner is the Bee fairly out of its waxen nursery, than the workers clear out the cell, and prepare it for the reception of honey. As soon as the cell is filled, the Bees close up the entrance with a waxen door, which is air-tight, and serves to preserve the honey in proper condition. Those who wish to eat honey in its pure state should always purchase it in the comb. If it be stored in pots, however well they may be sealed, it always crystallises, and in that state is injurious to digestion. Moreover, it is so extensively adulterated, that a pot of really pure honey is not easily obtained.

Besides the honey, 'bee-bread' is placed in the cells. This is a compound of honey and the pollen of flowers, and is chiefly used as food for the young grubs. We may often see the Bees hastening home with a load of yellow pollen on each of the hinder pair of legs, and this pollen is destined to be made into bee-bread.

Such, then, is a brief outline of the wonderful social habitation which is made by the Hive Bee.

WE now come to an insect which is as well known by name as the bee, though not so familiar to our eyes. This is the common HORNET (*Vespa crabro*), which is tolerably plentiful in many parts of England, but seems to be almost absent from others.

The nest of the Hornet is much like that of the wasp, except that it is proportionately larger, and is almost invariably built in hollow trees, deserted outhouses, and places of a similar description. Whenever the Hornet takes up its residence in an inhabited house, as is sometimes the case, the inmates are sure to be in arms against the insect, and with good reason. The Hornet is much larger than the wasp, and its sting is proportionately venomous. It is popularly said that three Hornets can kill a man; and although in such a case the sufferer must previously have been in bad health, the poisonous properties of the Hornet are sufficiently virulent to render such a saying popular.

Moreover, the Hornet is an irascible insect, and given to assault those whom it fancies are approaching its nest with evil intentions. It is not pleasant to be chased by wasps, but to be chased by Hornets is still less agreeable, as I can personally testify. They are so persevering in their attacks that they will follow a man for a wonderfully long distance, and if they be struck away over and over again, they will return to the charge as soon as they recover from the shock. There is a deep ominous menace in their hum, which speaks volumes to those who have some acquaintance with the language of insects; and no one who has once been chased by these insects will willingly run the same risk again.

HORNET.

Mr. S. Stone, whose interesting letter upon the wasp has already been mentioned, tells me that he has been successful in breeding Hornets as well as wasps, and forcing them to build nests much more beautiful than they would have made if they had been at liberty.

One nest, when of moderate size, was removed from the head of a tree, and placed in a large glazed box similar to those which have been mentioned in connection with the wasp. Within the box the Hornets continued their labours, and a most beautiful nest was produced, symmetrical in shape, and variegated with wonderfully rich colours. 'Such a nest as that,' writes Mr. Stone, 'is not produced by Hornets in a general way. They do not trouble themselves to form much of a covering, especially when a small cavity in the head of a tree is selected, which is often the case. The walls of the chamber they consider a sufficient protection for the combs

'If you expect them to form a substantial covering, the combs must be so placed as to have ample space around them, and if you expect them to fabricate a covering of great beauty, you must select the richest coloured woods, and such as form the most striking contrasts, and place them so that the insects shall be induced, nay, almost compelled, to use them in the construction of their nest. This is exactly what I did with reference to the nest in question.'

Knowing from experience the difficulty of assaulting a Hornet's nest, I asked Mr. Stone how he performed the task, and was told that his chief reliance was placed on chloroform. Approaching very cautiously to the nest, he twists some cotton wool round the end of a stick, soaks it in chloroform, and pushes it into the aperture. A mighty buzzing immediately arises, but is soon silenced by the chloroform, and as soon as this result has happened, mallet, chisel, and saw are at work, until the renewed buzzing tells that the warlike insects are recovering their senses, and will soon be able to use their formidable weapons. The chloroform is then re-applied until they are quieted, and the tools are again taken up.

The extrication of a nest from a hollow tree is necessarily a long and tedious process, on account of the frequent interruptions. Even if the insects did not interfere with the work, the labour of cutting a nest out of a tree is much harder than could be imagined by those who have not tried it.

Moreover, the habits of Hornets are not quite like those of

the wasps. At night, all the wasps retire into their nest, and in the dead of night the nest may be approached with perfect safety, the last stragglers having come home. Hornets are apt to continue their work through the greater part of the night, and if the moon be up, they are nearly sure to do so. Therefore, the nest-hunters are obliged to detail one of their party as a sentinel, whose sole business it is to watch for the Hornets that come dropping in at intervals, laden with building materials or food, and that would at once dash at the intruders upon their domains. Fortunately, the light from the lanterns seems to blind them, and they can be struck down as they fly to and fro in the glare.

The nest that has just been mentioned, was rather deeply imbedded in the tree, and cost no less than six hours of continuous labour, the work of excavation having been begun at eight P.M. and the nest extracted at two A.M. on the following morning.

In the illustration is seen a portion of a lately begun nest, much reduced in size, as may be conjectured from the dimensions of the insects that are crawling upon it. As the arrangement of the combs is identical with that of the wasp-nest, the interior is not disclosed. Another reason for showing the exterior of the nest is, that the reader might see how the Hornet forms the paper-like cover, and the manner in which the insects can enter at different parts, instead of having but a single entrance, as is the case with several hymenopterous nests which have been mentioned.

THERE is a very pretty, very interesting, and very destructive insect, called by entomologists the SMALL ERMINE MOTH (*Yponomeuta padella*), which is very plentiful in this country, and by gardeners is thought to be much too plentiful. It can easily be recognised by its long narrow wings, the upper pair of which are soft silvery, or satiny white, spotted with black, and the lower pair dark brown. The expanse of the spread wings is about three quarters of an inch.

In its winged and pupal states the insect is perfectly harmless, but in its larval condition it becomes a terrible pest.

Most caterpillars wage war singly on the foliage, and though they do much damage, their ravages are conducted in a desultory manner. The Small Ermines, however, band themselves together in hosts, and march like disciplined armies to the attack, invading a district and completely devastating it before they proceed to another.

They live in large tents, placed among the branches of some tree, and composed of silken threads, which are loosely crossed and recrossed in various directions. From this centre the caterpillars issue in vast numbers, each individual spinning a strong silken thread as it proceeds, which acts as a guide to the nest, just as the fabled clue led through the intricacies of Rosamond's bower. When once these caterpillars have taken possession of a tree, they are sure to strip it of its leaves as completely as if the foliage had been plucked out by hand. It is a very curious sight to watch the systematic manner in which these troublesome insects set about their work, how they send out pioneers which lead the way to new branches, either by crawling up to them or by lowering themselves to them by means of their silken life-lines, and how soon they are followed by their ever-hungry companions.

ANOTHER well-known British insect which constructs social habitations is the GOLD-TAILED MOTH (*Porthesia chrysorrhœa*), a familiar and beautiful insect, with wings of soft downy plumage, and snowy-white in colour, and a tuft of yellow hair at the end of the tail. The perfect insect may often be seen sticking on the trunks of trees in gardens, waiting until the evening, when it will fly off to its labours.

When the moth has laid its eggs, it plucks off the beautiful yellow tuft at the end of the tail, and with it forms a roof over the pile of eggs, laying the hairs so artificially as to make a perfect thatch. When the larvæ are hatched, they retain their sociability, and spin for themselves a common domicile. This house is very remarkable. Viewed on the exterior, it is seen to be a bag-like structure of whitish silk, rather strong and tough, but very yielding.

One of these nests, which I found in Wiltshire, is now before me. It was found in a hedge, about two feet from the ground, and is rather a complicated structure. The scaffolding, so to speak, of the nest is formed by a horizontal spray of three small twigs, and it is strengthened by the long hedge-grass which crossed the spray. Seeds of different kinds are woven into the walls, so that a comparatively small portion of the silk is exposed to view.

When cut open, it shows a singularly beautiful structure within. There are several sheets of silken tissue, each becoming more delicate, and the innermost being white, shining like satin; whereas the outer covering is dull-white, and very tough, clinging to the scissors so that a straight cut is almost impossible. Delicate walls divide the interior into several compartments, in all of which are evidences that the caterpillars must have resided for some time. The reason why the creatures make this nest is, that they are hatched towards the end of summer, and in consequence are forced to pass the winter in the larval condition, so that some warm residence is needful for them. It is well known that air is a very bad conductor of heat, and, in consequence, the successive sheets of silk which cover the nest, and which inclose layers of air between them, form a protection which is far warmer than would be obtained by a solid mass of silk measuring twice the thickness of the three walls, together with their intervening spaces.

THERE is an allied insect, popularly called the BROWN-TAILED MOTH (*Porthesia auriflua*), which spins a social nest that in many respects resembles that of the Gold-tailed Moth. The nest, however, is scarcely so elegant, nor is the silken web so beautifully delicate. Much, however, depends upon surrounding conditions, such as the disposition of the twig on which the nest is placed, and the presence or absence of leaves, whether those of the tree or of other plants that happen to grow in close proximity.

CHAPTER XXIV.

SOCIAL INSECTS—(*continued*).

A curious Ant from India (*Myrmica Kirbii*)—Locality of its nest—Description of the nest, its material and mode of structure—The DRIVER ANT of Africa—Description of the insect—Reason for its name—Its general habits—Destructive powers of the Driver Ant—How the insects devour meat and convey it home—How they kill snakes—Native legend of the python—Their mode of march—Fatal effects of the sunbeams—An extemporised arch—Method of escaping from floods—Site of their habitation—Modes of destroying them—Living ladders and their structure—Method of crossing streams—Tenacity of life—A decapitated Ant—Mode of biting—Description of the insect—AMAZON ANTS and their slaves—Curious nest of a Brazilian Wasp—Weight of the nest and method of attachment.

ALTHOUGH several species of Ants have been mentioned under the title of burrowing insects, there are many which possess very interesting habits, and which may here take their place among the creatures which build social habitations. Among them is a curious insect inhabiting India, and discovered by Colonel Sykes, the well-known naturalist, who called it *Myrmica Kirbii.*

This insect forms its nest on the branches of trees and shrubs, and Colonel Sykes mentions that he has found their curious habitation on the branches of the Kurwund shrub, *Carissa Carandas*, and on the Mango tree, *Mangifera Indica.*

The nests are more or less spherical, and are about as large as an ordinary foot-ball. The material of which they are made is cow-dung, which is spread in flakes in a manner that reminds the observer of the outside cover of a wasp's nest. The flakes are placed upon each other like the tiles of a house, so that although the insects can creep into the nest beneath the flakes, no water can enter. On the summit of the nest is one very large flake, that acts as a general roof to the structure.

Within the nest are placed a number of cells made of the same material as the exterior, and in them may be found insects in every state of development, eggs in one, larvæ in another, and pupæ in a third. No provision seems to be laid up within the nest, so that the inhabitants must depend on their daily excursions for their food.

The insects are extremely small, barely one-fifth of an inch in length, and are reddish in colour.

PERHAPS one of the most terrible of insects is that which is appropriately called the DRIVER ANT of Western Africa (*Anomma arcens*).

This insect is a truly remarkable creature. Although it is to be found in vast numbers, it has never been found in the winged condition, and neither the male nor the female have as yet been discovered. The workers are uniform in colour, but exceedingly variable in size. Their hue is deep brownish black, and their length varies from half an inch to one line, so that the largest workers nearly equal the common earwig, while the smallest are no larger than the familiar red ant of our gardens. In the British Museum are specimens of the workers, which form a regular gradation of size, from the largest to the smallest.

They are called Driver Ants, because they drive before them every living creature. There is not an animal that can withstand the Driver Ants. In their march, they carry destruction before them, and every beast knows instinctively that it must not cross their track. They have been known to destroy even the agile monkey, when their swarming hosts had once made a lodgment on its body, and when they enter a pigstye, they soon kill the imprisoned inhabitants, whose tough hides cannot protect them from the teeth of the Driver Ants. Fowls they destroy in numbers, killing in a single night all the inhabitants of the hen-roost, and having destroyed them, have a curious method of devouring them.

The Rev. Dr. Savage, who has experimented upon these formidable insects, killed a fowl and gave it to the Ants. At first, they did not seem to pay much attention to it, but he soon found that they were in reality making their preparations. Large parties of the insects were detached for the purpose of preparing

DRIVER ANTS.

a road, and worked with the assiduity which seems to be a characteristic of these energetic insects. Numbers of them were employed in smoothing the road to the nest by removing every obstacle out of the way, until by degrees a tolerably level road was obtained. The Ants are possessed of strength which seems gigantic when compared with their size, carrying away sticks four or five times as large as themselves, and never failing to pounce upon any grub or insect that might happen to be lurking beneath their shelter. They always carried such burdens longitudinally, grasping them with their jaws and legs, and passing the load under the body. Some of these roads are more than two hundred yards in length.

Meanwhile, the other Ants were busy with the fowl. Beginning at the base of the beak, they contrived to pull out the feathers one by one, until they stripped it regularly backwards, working over the head, along the neck, and so on to the body. This was evidently a very hard task, as the insects did not possess sufficient strength to pull out the feathers by main force, and were consequently obliged to grub them up laboriously by the roots. The next business was to pull the bird to pieces, and at this work they were left. Unfortunately the experiment was spoiled by the natives, who stole the fowl, thinking that the Ants had eaten so many of their poultry that they were justified in retaliation. Others chose to excuse themselves by saying that they thought the fowl to be a fetish offering to the Ants, and accordingly took it away from them.

The large iguana lizards fall victims to the Driver Ants, and so do all reptiles, not excluding snakes. It seems, from the personal observations of Dr. Savage, that the Ants commence their attack on the snake by biting its eyes, and so blinding the poor reptile, which only flounders and writhes helplessly on one spot, instead of gliding away to a distance.

It is said by the natives, that when the great python has crushed its prey in its terrible folds, it does not devour it at once, but makes a large circuit, at least a mile in diameter, in order to see whether an army of Driver Ants is on the march. If so, it glides off, and abandons its prey, which will soon be devoured by the Ants; but if the ground is clear, it returns to the crushed animal, swallows it, and gives itself to repose until the process of digestion be completed. Whether this assertion be true or not, Dr. Savage cannot say; but it is here given in order to show the extreme awe in which the natives hold the Driver Ants.

So completely is the dread of them on every living creature, that on their approach whole villages are deserted, and in extreme cases the entire population is forced to take to the rivers, knowing that the insects will not enter water unless obliged to do so; although on occasion they do not hesitate to commit themselves to the waves, as will presently be seen.

The order of their marching is very curious, and is well described by Dr. Savage:—

'Their sallies are made in cloudy days, and in the night, chiefly in the latter. This is owing to the uncongenial influence of the sun, an exposure to the direct rays of which, especially when the power is increased by reflection, is almost instantaneously fatal. If they should be detained abroad till late in the morning of a sunny day by the quantity of their prey, they will construct arches over their path, of dirt agglutinated by a fluid excreted from their mouth. If their way should run under thick grass, sticks, &c., affording sufficient shelter, the arch is dispensed with; if not, so much dirt is added as is necessary to eke out the arch in connection with them. In the rainy season, or in a succession of cloudy days, the arch is seldom visible; their path, however, is very distinct, presenting a beaten appearance, and freedom from everything moveable.

'They are evidently economists in time and labour; for if a crevice, fissure in the ground, passage under stones, &c., come in their way, they will adopt them as a substitute for the arch.

'In cloudy days, when on their predatory excursions, or migrating, an arch for the protection of the workers is constructed of the bodies of their largest class. Their widely-extended jaws, long slender limbs, and projecting antennæ, intertwining, form a sort of net-work, that seems to answer well their object. Whenever an alarm is given, the arch is instantly broken, and the ants, joining others of the same class on the outside of the line, who seem to be acting as commanders, guides, and scouts, run about in a furious manner, in pursuit of the enemy. If the alarm should prove to be without foundation, the victory won, or danger passed, the arch is quickly renewed, and the main column marches forward as before, in all the order of an intellectual military discipline.'

Sometimes, as is usual in tropical countries, the rain descends like a flood, converting in a few minutes whole tracts of country into a temporary lake. The dwellings of the Driver Ant are immediately deluged, and, but for a remarkable instinct which is implanted in the insects, most of the Ants, and all the future brood, would perish. As soon as the water encroaches upon their premises, they run together and agglomerate themselves into balls, the weakest (or the 'women and children,' as the natives call them) being in the middle, and the large and powerful insects on the outside. These balls are much lighter than water, and consequently float on the surface, until the floods retire and the insects can resume their place on dry land.

The size of the ant-balls is various; but they are, on an average, as large as a full-sized cricket-ball. One of these curious balls was cleverly caught in a handkerchief, put in a vessel, and sent to Mr. F. Smith, of the British Museum, who has kindly presented me with several specimens of the insect.

When a colony of these insects has been established near a house, the inhabitants naturally endeavour to destroy it. The habitation is very simple and artless, and generally consists of a mere hole in a rock or bank, in which the creatures assemble. They are very fond of usurping the sepulchres of the dead, which are usually excavated in the sides of hills, and are about eighteen inches in depth.

The natives generally try to destroy the colony by heaping dry leaves of the palm upon the dwelling, and setting fire to the heap. When this plan was tried, it was found to be very unsatisfactory; for the greater mass of the insects contrived to make their escape, and were found upon neighbouring trees, clinging in heavy bunches and long festoons, which connected one branch with another, and formed ladders over which the insects could pass. These festoons were made in a very curious manner.

First, a single Ant clung tightly to a branch, and then a second insect crawled cautiously down its suspended body, and hung to its long, outstretched limbs. Others followed in rapid succession, until they had formed a complete chain of Ants, which swung about in the wind. One of the largest workers then took its stand immediately below the chain, held firmly to the branch with its hind limbs, and dexterously caught with its fore-legs the end of the living chain as it swung past. The ladder was thus completed, and fixed ready for the transit of insects; and, in a similar way, the whole tree was covered with festoons of Ants, until it was blackened with their sable bodies.

They can even cross streams by means of these ladders. Crawling to the end of a bough which overhangs the water, they form themselves into a living chain, and add to its length until the lowermost reaches the water. The long, wide spread-limbs of the insect can sustain it upon the water, especially when aided by its hold on the suspended comrade above.

Ant after Ant pushes forward, and the floating portion of the chain is thus lengthened, until the free end is swept by the stream against the opposite bank. The Ant which forms the extremity of the chain then clings to a stick, stone, or root, and grasps it so firmly, that the chain is held tightly, and the Ants can pass over their companions as over a suspension bridge. In the illustration a column of Driver Ants is shown on the march. The vanguard of the column has crossed the stream by means of the living ladder, which is seen suspended from a branch, and extended across the water. The fragile tube which they build is also shown, and a few of the larger architects are drawn of the natural size. The smaller specimens will not emerge from the tunnel.

In Dr. Livingstone's well-known work, there are several interesting accounts of ants and their habits, and one anecdote bears so aptly on the subject, that I give it in the writer's own words.

After describing the terrible drought at Chonuane, when the river Kolobay ran dry and the fish perished, when the crocodile himself was stranded and died, and the native trees could not hold up their leaves, he proceeds as follows:—'In the midst of this dreary drought, it was wonderful to see those tiny creatures, the Ants, running about with their accustomed vivacity. I put the bulb of a thermometer three inches under the soil in the sun at mid-day, and found the mercury to stand at 132° to 134°; and if certain beetles were placed on the surface, they only ran about a few seconds and expired.

'But this boiling heat only augmented the activity of the long-legged Black Ants; they never tire; their organs of motion seem endowed with the same power as is ascribed by physiologists to the muscles of the human heart, by which that part of the frame never becomes fatigued, and which may be imparted to all our organs in that higher sphere to which we fondly hope to rise.

'Where do these Ants get their moisture? Our house was built on a hard, ferruginous conglomerate, in order to be out of the way of the White Ant, but they came despite the precaution; and not only were they in this sultry weather able individually to moisten soil to the consistency of mortar for the formation of galleries, which in their way of working is done by night (so that they are screened from the observation of birds by day in passing and repassing towards any vegetable matter they may wish to devour), but, when their inner chambers were laid open, these were also surprisingly humid; yet there was no dew, and the house being placed on a rock, they could have no subterranean passage to the bed of the river, which ran about three hundred yards below the hill. Can it be that they have the power of combining the oxygen and hydrogen of their vegetable food by vital force as to form water?'

Three species of Driver Ant are known, namely, the common species, which has already been described, *Anomma Burmeisteri*, and a smaller species, *Anomma rubella*.

The two first insects are deep, shining black, and resemble each other so closely that an unpractised eye could not distinguish between them, while the last may be easily known by its brownish red hue.

The specimens which have already been mentioned are now

before me, and curious beings they are. The largest are black, with a slight tinge of red, and have an enormous head, almost equalling one-third of the entire length. It is deep and wide as well as long, as indeed is necessary for the attachment of the muscles which move the enormous jaws. These weapons are sharply curved, and when closed, they cross each other, so that when the insect has fairly fixed itself, its hold cannot be loosened unless the jaws are opened. It is useless, therefore, to kill the ant, for its head will retain its grasp in death as well as in life. Beside the sharp point of the mandibles, they are further armed with a central tooth, which is so formed that when the mandibles are quite closed, and the points crossed to the utmost, the tips of the central teeth meet and form another means of grasping.

There is no vestige of external eyes, and even the half-inch power of the microscope fails to show the slightest indication of visual organs. As, however, the horny coat of the head is sufficiently translucent to permit the articulation of the jaws to be seen through it, when a very powerful light is thrown upon the head, and the eyes of the observer are well sheltered, it is possible that the insect may have some sense of sight, and at all events will be able to distinguish light from darkness.

The limbs are of a paler red than the body, and although they are slender and delicate, their grasping power is very great. Two of my specimens had grasped each other's limbs with such force that they could not be separated without damaging the insect, and it was not until the rigid joints were softened with moisture, and then with the aid of a magnifier, that I succeeded in disengaging the insects.

The smaller specimens are not so black as the larger, nor are their jaws so proportionately large, but they are still formidable insects, if not from their individual size, yet from their collective numbers and their reckless courage, which urges them to attack anything that opposes them. Fire will frighten almost any creature, but it has no terrors for the Driver Ant, which will dash at a glowing coal, fix its jaws in the burning mass, and straightway shrivel up in the heat.

The remarkable fact has already been mentioned, that two species of Wasp will inhabit the same nest, and amicably work at the same edifice. Entomologists have long been aware that two species of Ant will dwell in the same nest, and live upon friendly terms, although the association of the working part of the community is not voluntary, but compulsory.

The Ant which employs enforced labour is called the Amazon Ant (*Polyergus rufescens*), and is tolerably common on the Continent. This insect is not furnished with jaws which are capable of performing the work that usually falls to the lot of the neuters; but the same length and sharpness of the mandibles which unfit the insect for work, render it eminently capable of warfare. When, therefore, a colony of the Amazon Ants is about to establish itself, the insects form themselves into an army, and set off on a slave-hunting expedition.

There are at least two species of Ant which act as servants to the Amazon Ants, the one being named *Formica fusca*, and the other *Formica cunicularia*; and to the nests of one or other of these insects the Amazons direct their march.

As soon as they reach the nest, they penetrate into all its recesses, in spite of opposition, and search every corner for their spoil. This consists solely of the pupæ which will afterwards be developed into neuters; and vast numbers of the unconscious young are carried off in the jaws of the conquerors. The rightful owners and relatives of the captured young cannot resist the enemy, as their shorter though more generally useful jaws are unable to contend with the long and sharply-pointed weapons of their foes.

After the marauding army has returned, the living spoils are carefully deposited in the nest, where they are speedily hatched into perfect insects of the worker class, and immediately take on themselves the labours of the nest, just as they would have done in their own home. The Amazon Ant seems to be utterly incapable of work; and in one notable instance, when a number of them were confined in a glass-case, together with some pupæ, they were not only unable to rear the young, but could not even feed themselves, so that the greater number died from hunger. By way of experiment, a single specimen of the slave Ant (*Formica fusca*) was introduced into the case, when the state of affairs was at once altered. The tiny creature undertook the whole care of the family, fed the still living Amazon Ants, and took charge of the pupæ until they were developed into perfect insects.

Some writers have enlarged upon the hard lot of the slave Ants, imagining their servitude to be as distasteful to them as it is sometimes made to human slaves. Mr. Westwood, however, points out very clearly that any compassion bestowed upon them is wasted, and that the lot of the 'helots'—if they may be so called—is precisely that for which they were made. The labours which the little creatures undertake are not arbitrarily forced upon them by the dread of punishment, but are urged upon them by the instincts implanted within them. They would have worked in precisely the same manner and with exactly the same assiduity, in their own nests as in that of their captors, and the labours are undertaken as willingly in the one case as in the other.

They find themselves perfectly at home, and are in every respect on a par with their so-called masters. In point of fact, however, the real masters in the nest are the slaves, for upon them the Amazons are dependent from their earliest days to the end of their life, and without them the entire community would perish. The slaves have no other home but that to which they have been brought, and are no more to be pitied than are dogs, cattle, and other domestic animals that never have freedom. Indeed, none but solitary animals can be free even in the wild state, for they are held in absolute servitude by the leaders of the herds, and, if they dare to disobey, are summarily punished.

As the slaves are always neuters, it is necessary that fresh importations should be made as fast as the demand for workers exceeds the supply; and it is really a wonderful thing that the

Amazon Ants should always select the pupæ which will afterwards be developed into neuters, and never take those from which males or females will issue.

The Amazon of the Continent is not the only Ant which enslaves the neuters of another species, for in different parts of the world several species of Ants have been observed which seize upon workers belonging to other nests, and bring them to do the work of the home. A Brazilian species (*Myrmica paleata*) has been observed to act in a similar manner.

In the collection of the British Museum may be seen a very remarkable nest; which is made by some species of wasp at present unknown.

The material of which it is formed is mud, or clay, which is kneaded by the insect until it has attained a wonderful tenacity and strength, and is rendered so plastic as to be worked almost as neatly as the waxen bee-cell. It is of rather a large size, measuring about thirteen inches in length, by nine in width, and filled with combs. Unfortunately, in its passage to this country, it was broken and much damaged, but the fragments were collected and skilfully put together by Mr. F. Smith, who has succeeded in restoring the nest to its original shape, with the exception of an aperture through which the interior of the nest may be seen.

The accident was in so far an advantage, that it gave opportunities of studying the construction of a nest which is at present unique, and which the officers of the Museum might be chary of cutting open, particularly as its materials are so brittle. The walls of the nest are remarkably hard and solid, but extremely variable in thickness, some parts being nearly three times as strong as others. The upper portions of the nest are the thickest, the reason for which is evident on inspecting the specimen.

The nest was found in a Guianan forest, near the river Berbice, suspended to a branch, which passed through a hole in the solid wall of the nest. In the actual specimen, the branch is wanting; but in the illustration it has been restored, in order to show the manner in which the winged artificers suspended their wonderful home. As is always the case with pensile nests, the foundation is laid at the top, thus carrying out Dean Swift's suggestion for a new patent in architecture.

MUD WASP.

A large quantity of clay is worked round the chosen branch, and made very strong, in order to sustain the heavy weight which will be suspended from it. This clay foundation is wonderfully hard, though very brittle, this latter quality being probably due to the long residence in a room which is always kept warm and dry by artificial means. In the open air, and in the ever damp, though hot atmosphere of tropical America, the clay would probably be much tougher, without losing the necessary hardness.

The combs are not flat, like those of an ordinary wasp-nest, but are very much curved, so that when the nest is laid open they almost follow the curve of the walls. This peculiar form of the comb is shown in the illustration. The cells are not very large, scarcely equalling the worker cells of the common burrowing wasp of England.

One of the most remarkable points in the construction of this nest is the entrance. In pensile nests, the insect usually forms the opening below, so that it may be sheltered from the wind and rain. Moreover, it is usually of small dimensions, evidently in order to prevent the inroads of parasitic insects and other foes, and to give the sentinels a small gateway to defend. But the particular Wasp which built this remarkable nest seems to have set every rule at defiance, and to have shown an entire contempt of foes and indifference to rain.

As may be seen by reference to the illustration, the entrance is extremely long, though not wide, and extends through nearly the length of the nest, so that the edges of the combs can be seen by looking into the aperture. The edges of the entrance are rounded, so that the outer edge is wider than the inner; but it is still sufficiently wide to allow the little finger of a man's hand to be passed into the interior; while its length is so great, that forty or fifty insects might enter or leave the nest together.

CHAPTER XXV.

PARASITIC NESTS.

Various Parasites—Parasitic Birds—The CUCKOO and its kin—The COW BIRD and its nest—Size of its egg—The BLUE-FACED HONEY-EATER or BATIKIN—General habits of the bird—Singular mode of nesting—The SPARROW-HAWK and its parasitic habits—The KESTREL, its quarrel with a Magpie—The STARLING and the Pigeons—The PURPLE GRAKLE or CROW BLACK-BIRD—Its curious alliance with the Osprey—Wilson's account of the two birds—Parasitic Insects—The ICHNEUMON FLIES—The parasite of the CABBAGE CATERPILLAR—Its numbers and mode of making its habitation—Trap-doors of the cells—The Australian Cocoon and its parasites—The OAK-EGGER MOTH, its cocoons and enemies—RUBY-TAILED FLIES and their victims—Modes of usurpation—The CUCKOO FLIES or Tachinæ—Parasites on vegetables—The GALL FLIES and their home—British Galls, their shapes, structures, and authors.

WE now pass to another branch of this inexhaustible subject, and come to those creatures that are indebted to other beings for their homes. In some cases, the habitation is simply usurped from the rightful proprietors, who are either driven out by main force or are ousted by gradual encroachment. In other cases, the deserted tenement of one animal is seized upon by another, which either inhabits it at once, or makes a few alterations, and so converts it to its own purposes. In many instances, however, the habitation of the parasite is found within the animal itself; and in some cases the entire body forms the home of the parasite.

The kingfisher, for instance, usurps the deserted hole of a water-shrew; and the humble-bee and wasp usually take advantage of the deserted burrow of some rat or mouse. In the account of the sociable weaver-bird, mention is also made of certain little green parrots, which are apt to take possession of the great nest, and use it for their own purpose. And in the last chapter an example was mentioned where a carder-bee established herself in the deserted nest of a wren, and so saved herself the trouble of fetching materials and building a dome.

Birds of various kinds are notorious parasites, the Cuckoos ranking as chief among them, inasmuch as they make no nest at all, but simply lay their eggs in the nests of other birds, and foist upon them a supposititious offspring, which occupies the entire nest and monopolises all the care of its foster-parents.

All Cuckoos, however, do not possess this habit; for some of the group build nests which are remarkable for their beauty, and tend their young as carefully as do any birds. The celebrated Honey-finders, for example, which are found in most hot portions of the globe, are notable for their skill in architecture. The nests of these birds are pensile, and not unlike those of the African weaver-birds, which have already been described. They are made of tough bark, torn into filaments, and are flask-like in shape, hung from the branches of trees, and having their entrance from below.

Then there is the well-known Cow-bird of America (*Coccygus Americanus*), which is closely allied to the common cuckoo, and yet which builds its own nest, and rears its own young. 'Early in May,' writes Wilson, 'they begin to pair, when obstinate battles take place among the males. About the 10th of that month they commence building. The nest is usually fixed among the horizontal branches of an apple-tree; sometimes in a solitary thorn, crab, or cedar, in some retired part of the woods. It is constructed with little art, and scarcely any concavity, of small sticks and twigs, intermixed with green weeds and blossoms of the common maple. On this almost flat bed the eggs, usually three or four in number, are placed; these are of an uniform greenish blue colour, and of a size proportionate to that of the bird.

'While the female is sitting, the male is generally not very far distant, and gives the alarm by his notes, when any person is approaching. The female sits so close, that you may almost reach her with your hand, and then precipitates herself to the ground, feigning lameness, fluttering, trailing her wings, and tumbling over, in the manner of the partridge, woodcock, and many other species. Both parents unite in providing food for the young.'

In Australia there is a large group of rather pretty birds, popularly called Honey-eaters, because they feed largely on the sweet juices of many flowers, although the staple of their diet consists of insects. They seem indeed to occupy in Australia the position which is taken in America by the humming-birds, and by the sun birds of the old world. To this group belong many familiar and interesting species, such as that which produces a sound like the tinkling of a bell, and is in consequence called the Bell-bird; the different species of Wattle Birds; the odd, bald-headed Friar Birds, and the splendidly decorated Poe Birds.

One species of it, which comes in the present section, is the Blue-faced Honey-eater of New South Wales, called by the natives Batikin (*Entomyza cyanotis*). It is a pretty bird, the plumage being marked boldly with black and white, and a patch of bare skin round the eyes being bright azure. This peculiarity has earned for the bird the specific title of *cyanotis*, or 'blue-eared.'

Like all the Honey-eaters, it is a most lively and interesting bird, and to the careful observer affords an endless fund of amusement. It is never still, but traverses the branches with astonishing celerity, skipping from one to another, probing every crevice with its needle-like tongue, hanging with its head downwards, and even suspending itself by a single claw, while it secures a tempting insect. It is generally to be found on the eucalypti, or gum-trees, and is one of the stationary birds, remaining in the same locality throughout the year.

The generality of the Honey-eaters are skilful architects, but the Batikin seems not to share the ability of its relatives, or, at all events, not to exercise it. Mr. Gould thinks that the bird can hardly depart so far from usual custom as to be incapable of building a nest, but he has never found such a nest, nor heard of one. The Batikin is one of the parasitic group, usurping the nest of another bird, and taking possession of it in a very curious fashion.

In Australia there is a bird belonging to the genus *Pomatorhinus*, which somewhat resembles the bee-eater, except in plumage, which is quite dull and sober. This bird builds a large, domed edifice, and appears to make a new nest every year. The deserted nests are always usurped by the Batikin, which establishes herself without any trouble. The reader would naturally imagine that when the bird finds herself in possession of so large and warm a nest, she will pass into the interior, and hatch her young under the protection of the roof. This plan, however, she does not follow, preferring to take up her abode on the very top of the nest, exposed to all the elements. She takes very little trouble about preparing her home, but merely works a suitable depression upon the soft dome, lays her eggs in it, and there hatches them.

The reader will remember that there are several birds which form a supplementary nest upon the exterior of the original domicile, and the parasitic nest of the Batikin is evidently an extension of the same principle.

In England we have many parasite birds, one of which is the common Sparrow-Hawk (*Accipiter Nisus*), which is in the habit of usurping the nest of the common crow, magpie, or other bird, and laying its handsome eggs therein.

Whether it forcibly drives away the rightful owner, or whether it contents itself with a nest which has already been abandoned, is not precisely known, different naturalists inclining to opposite opinions. In all probability, therefore, both disputants are right, and the Sparrow-Hawk takes a deserted nest when it can find one, and when it cannot do so, attacks birds which are in actual possession of a suitable nest, and takes possession of their home. In such a case, the combat must be a sharp one, for both crow and magpie are courageous birds, nothing inferior in determination to their assailant, and armed with bills which are much larger, and quite as formidable as that of the Sparrow-Hawk.

The KESTREL (*Tinnunculus alaudarius*) is also in the habit of laying its eggs in the nest of other birds, and may possibly eject the rightful owner by main force. This opinion is rendered probable by a fact mentioned by Mr. Peachey, in the 'Zoologist.' A man was passing a tree, and hearing a loud screaming proceeding from a nest at the summit, he had the curiosity to climb the tree. The screams still continued, and on putting his hand into the nest, he found two birds struggling, the uppermost of which he caught. This proved to be a Kestrel, and as soon as it was secured, the other bird, which was a magpie, flew out, evidently having been worsted by its antagonist.

Then there is the well-known STARLING (*Sturnus vulgaris*), which is a notably parasitic bird, delighting to take the nests of the jackdaw, pigeon, and other birds, and to use them as its own. Every one who has a dovecote knows how apt are the Starlings to usurp the boxes intended for the pigeons, and how in consequence it is accused of killing the young of the pigeons, and sucking their eggs, two accusations which I believe to be wholly false. Were the Starlings to be thus predacious, the pigeons would be quite aware of their depredations, and would appear greatly disturbed whenever the robbers were seen. As, however, the pigeons in one box live in perfect amity with the Starlings in the next, it is very unlikely that the latter birds prey in any way upon the former.

THERE is a group of birds which are popularly called Grakles, and are scientifically known as Quiscalinæ. They are also called Boat-tails, because their tail-feathers are formed so as to take the shape of a canoe. One species, the PURPLE GRAKLE, or CROW-BLACKBIRD (*Quiscalus versicolor*), is conspicuous as a parasitic bird, and selects a most extraordinary spot for its nest.

Generally, the predacious birds are avoided and feared by the rest of the feathered tribes, and if a hawk or eagle show itself, the smaller birds either hide themselves, or try to drive away the intruder by force of numbers or swiftness of wing. The

Purple Grakle, however, is devoid of such fears, at all events as far as one species of predacious bird is concerned, and boldly takes up its abode with the osprey or fish-hawk (*Pandion haliaetus*).

The nest of the osprey is a very large edifice, made of sticks, grass, seaweed, leaves, and similar materials. The foundations are made by sticks almost as thick as broom-handles, and some two or three feet in length, on which are piled smaller sticks, until a heap some four or five feet in height is made. Interwoven with the sticks are stalks of corn and various herbs, the larger seaweeds and large pieces of grass, the whole mass being a good load for an ordinary cart, and as much as a horse can be reasonably expected to draw.

As the sticks of which the foundation of the nest are made are very large, and not regular in form, considerable interstices are left between them, and in such spots the Grakle chooses to nidificate.

In writing of the osprey, Wilson remarks as follows : 'There is one singular trait in the character of this bird which is mentioned in treating of the Purple Grakle, and which I have had many opportunities of witnessing. The Grakles, or Crow-Blackbirds, are permitted by the fish-hawk to build their nests among the interstices of the sticks of which its own is constructed,—several pairs of Grakles taking up their abode there like humble vassals around the castle of their chief,—laying, hatching their young, and living together in mutual harmony. I have found no less than four of these nests clustered round the sides of the former, and a fifth fixed on the nearest branch of the adjoining tree, as if the proprietor of this last, unable to find an unoccupied corner on the premises, had been anxious to share, as much as possible, the company and protection of this generous bird.' In another place, the same writer remarks that the curious allies 'mutually watch and protect each other's property from depredators.'

The Purple Grakle is, however, perfectly capable of building a nest for itself. Indeed, the generality of the birds build in tall trees, usually associating together, so that fifteen or twenty nests are made in the same tree. The nests are well and carefully made of mud, roots, and grasses, about four inches in depth, and warmly lined with horsehair and very fine grasses. The fact that the bird possesses this capability of nest-building, gives more interest to the occasional habit of sharing its home with the osprey—a privilege of which it seems to avail itself whenever an osprey's nest is within reach.

The colour of this bird appears at a little distance to be black, but is in reality a very deep purple, changing in different lights to green, violet, and copper, and having a glossy sheen like that of satin.

We now pass to the Parasitic Insects. As this work is intended to describe dwellings which are in some way formed by the creatures, it is necessary to exclude all the parasite insects that may exist upon the animal, and make no habitation, such as the ticks, as well as those which are merely parasitic within the animal, such as the various entozoa.

Of Parasitic Insects, the greater number belong to that group of hymenoptera which is called Ichneumonidæ, and which embraces a number of species equal to all the other groups of the same order. Being desirous of producing, as far as possible, those examples of insects which have not been figured, I have selected for illustration several specimens which are now in the British Museum, one or two of which have only been recently placed in that collection.

The best known of all the Ichneumonidæ is that tiny creature called *Microgaster glomeratus*.

A group of these insects and their cells is now before me, and will be briefly described.

Small as it is, this tiny insect is extremely valuable to us, and to the gardener is beyond all value, though, as a general rule, the gardener knows nothing about it. Where it not for this ichneumon, we should scarcely have a cabbage or a cauliflower in the garden; for the noisome cabbage caterpillars would destroy every leaf of the present plant, and nip the growth of every bud which gave promise for the future.

Every one knows the peculiarly offensive caterpillars which eat the cabbages, and which are the offspring of the common large white butterfly. In the spring, the butterflies may be seen flitting about the gardens, settling on the cabbages for a few moments, and then flying off again. They look very pretty, harmless creatures, but, in fact, they are doing all the harm that lies in their power. Forty or fifty eggs are thus laid on a plant,

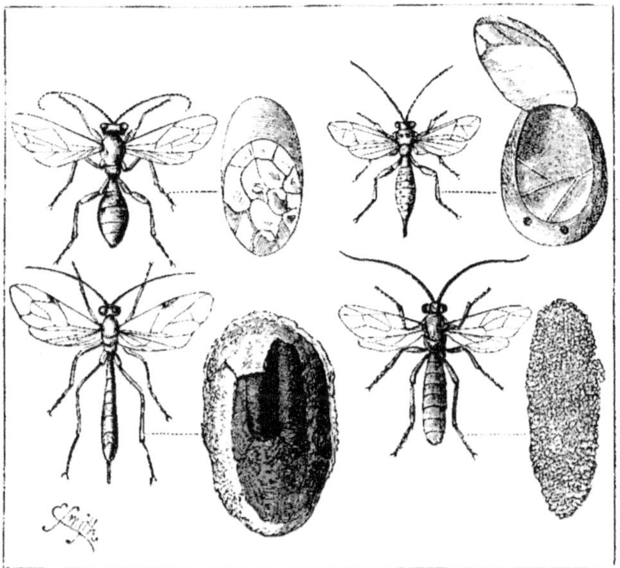

PARASITIC INSECTS.

COCOON OF OAK-EGGER MOTH. (*Cryptus fumipennis*)

COCOON FROM NEW SOUTH WALES. (*Pimpla*)

COCOON OF PUSS MOTH. (*Paniscus glaucopterus.*)

COCOON OF GOAT MOTH. (*Lamprosa setosa.*)

and if only one quarter of the number are hatched, they are quite capable of marring every leaf. In process of time, they burst from the egg-shell, and commence their business of eating, which is carried on without cessation throughout the whole time of the larval existence, with a few short intervals, while they change their skins.

When they are full grown, they crawl away from the plant to some retired spot, and there suspend themselves, preparatory to changing into the pupal condition. A few of them succeed in this task, but the greater number never achieve the feat, having been the unwilling nourishers of the ichneumon flies. Just before the larva is about to pass into the pupal state, a number of whitish grubs burst from its sides, and each immediately sets to work at spinning a little yellow, oval cocoon. The walls of the cocoon are hard and smooth, especially in the interior; but the outside is covered with loose floss-silk, which serves to bind all the cocoons together. Generally, they are very loosely connected; but a group of these little objects is now before me, where the cocoons are formed into a flattish oval mass, about the size and shape of a scarlet-runner bean, split longitudinally, and are bound so tightly together, that their shape can barely be distinguished through the enveloping threads.

As is the case with the cells of the Burnet ichneumon, each cell is furnished with a little circular door which exactly resembles in shape and dimensions the circular pieces of paper that are punched out of the edges of postage-stamps. On the average, about sixty or seventy ichneumon flies are produced from a single cabbage caterpillar.

The groups of yellow cells are very plentiful towards the middle of summer and the beginning of autumn, and may be found on walls, palings, the trunks of trees, in outhouses, and, in fact, in every place which affords shelter to the caterpillar. Nothing is easier than to procure the insects from the cocoons, as the yellow mass needs only to be put into a box, with a piece of gauze tied over it by way of a cover. Nearly every cocoon will produce its ichneumon, and as the little creatures are not strong-jawed enough to bite through the gauze, they can all be secured.

There are many species of Microgaster; but those which have been mentioned are the most important, and make the most interesting habitations.

THE large oval cocoon was brought from New South Wales,

and is evidently the produce of some lepidopterous insect, probably a moth allied to the silkworm. Upon the larva which constructed the cocoon an ichneumon has laid her eggs, and the consequence has been that the caterpillar has been unable to change into the pupal condition, but has succumbed to the parasites which infested it. These insects are not of minute dimensions, like the Microgaster, but are tolerably large, and in consequence can be but few in number. The cells are very irregular in shape, and are not rounded like those of many Ichneumonidæ, but have angular edges.

WITHIN the same case there are several cocoons in which a similar calamity has befallen the caterpillars which made them. There is, for example, a cocoon of the OAK-EGGER MOTH (*Lasiocampa quercus*), the interior of which resembles that of the insect which has just been described, except that the cells of the parasite are more numerous. This species of caterpillar is peculiarly subject to the attacks of the ichneumon flies, as is well known to all practical entomologists, who lose many of their carefully bred specimens by means of these insects.

There is also one of the winter cocoons of the GOAT MOTH caterpillar, the inmate of which has been pierced by the ichneumon fly, and killed by its young. As the species of ichneumon is a large one, only a single individual was produced, and as may be seen from the cell of the parasite which is placed by the side of its victim, the habitation of the ichneumon is so large that it must have occupied nearly the entire cocoon of the dead caterpillar.

THOSE splendid insects which are popularly called RUBY-TAILED FLIES, or FIRETAILS, and scientifically are termed *Chrysididæ,* are also to be numbered among the parasitic insects.

They make no nests for themselves, but intrude upon those of various mason and mining bees, and several other insects. The Firetail does not, however, lays its eggs in the body of the larva, but makes its way into the nest while the rightful owner is absent, and places an egg near that of the bee. The egg of the parasite is sometimes hatched at the same time with that of the bee, but generally later. In the first instance, the larva feeds on the provisions which were supplied for the bee, and so starves the poor creature to death ; and in the latter case, it is not hatched until the young bee is large and fat, and capable of affording ample subsistence to the parasite, which fastens upon it and devours all the softer portions.

Then there are the CUCKOO FLIES (*Tachinæ*), which bear some resemblance to the common house-fly, but which are parasitic, feeding on the larvæ of other insects, and selecting the same species which are persecuted by the firetails. When the Tachina larva has eaten that of the mason bee, it forms an oval cocoon, and there remains until the time for becoming a perfect insect. A single larva of the mason bee seems to be sufficient for the Tachina grub, as Mr. Rennie has recorded an instance where two larvæ of the mason bee were in a nest into which a single egg of a Tachina had been introduced. The parasitic larva devoured one of the rightful inhabitants, but did not touch the other, and the cocoons of the bee and the Tachina were formed side by side.

WE now pass to a remarkable series of insects belonging to the same order as the ichneumons, but parasitic upon vegetables and not on animals. Their scientific name is *Cynipidæ*, and they are popularly known as GALL FLIES, because they cause those singular excrescences which are so familiar to us under the name of Galls. This group comprises a vast number of species, all of which have a strong family resemblance, though they greatly differ from each other in size, form, and colour.

In the accompanying illustration are given several examples of British Galls, most of which are tolerably common in this country, and some of which can be found in plenty.

IN the left hand upper corner of the illustration is a figure of an oak-leaf, upon which are two globular projections. These are the well-known 'cherry-galls,' which are made by a little insect called *Cynips quercus-folii.* They are beautifully coloured, some being entirely scarlet, while others are white, orange, and red, in various gradations, something like the colour of a nearly ripe peach, or those of a Newtown pippin. Perhaps they bear more resemblance to the apple than to the peach, because their

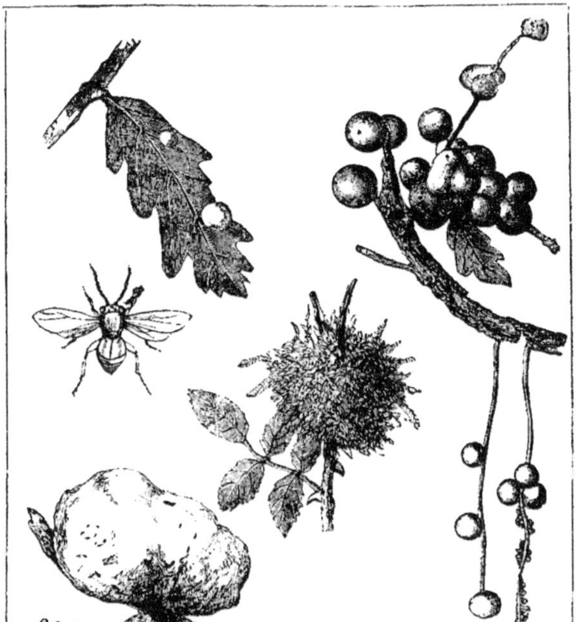

BRITISH GALLS.

Leaf Galls of Oak. Bedeguar of Rose. Galls of Cynips Kollari.
Cynips Kollari (slightly magnified). Currant Galls of Oak.
Oak Apple.

surface is highly polished and shining, much like that of the American apple.

These galls may be found in profusion upon the oak-leaves, and are most plentiful upon pollard oaks, upon the youngest trees, or upon the oak underwood that sprouts around a felled trunk.

If one of the galls be cut open with a knife, it will be found to consist of a soft, pulpy substance, fuller of juice than an apple, and somewhat resembling the consistence of a hothouse grape. In the very centre of the soft mass the knife will meet with resistance in the shape of a globular cell of hard, woody texture, and in the middle of the cell will be found a tiny grub, perfectly white, very fat, somewhat resembling the grub of the humble bee, and curved so as to fit the globular cell in which it lies. This is the little being for whose benefit the gall was formed, and the little white grub feeds on the juices of the gall, precisely as the larva of the ichneumon fly feeds on the soft portions of the insect in which it temporarily resides.

On seeing the little creature thus snugly ensconced in the receptacle which serves it at once for board and lodging, a question naturally arises as to the manner in which it was placed there. No aperture is perceptible in the gall, not a hole through which air can reach the enclosed larva, which must, therefore, be capable of existing without more air than can pass through the minute pores of the vegetable substance in which it lies, or must be able to respire by means of the oxygen which is given out by living plants.

The question, indeed, is very like the well-known query as to the manner in which a model of a waggon and four horses can find its way into a bottle, the neck of which is so small as to prevent even the head of the waggoner from passing. The answer is similar in both cases. The bottle was ingeniously blown over the waggon and horses, and the gall was formed around the grub.

When its leaf is in its full juiciness, and the sap is coursing freely through its textures, a little black insect comes and settles upon the leaf. She is scarcely as large as a garden ant, but has four powerful and handsome wings, which can be used with much agility. An entomologist, on seeing her, would at once pronounce her to belong to the order hymenoptera, and to be closely allied to the ichneumon flies which have just been described.

Running to and fro upon the leaf, she fixes upon one of the nervures, and there remains for a short time, evidently busy about some task, which is very important to her, but which her minute size renders impossible to be observed with the naked eye. If, however, a magnifying glass be applied very carefully to the leaf, the following process will be seen.

From the abdomen there projects a tiny hair-like ovipositor, which is coiled in such a manner that it can be protruded to a considerable length. This ovipositor is thrust into the leaf, so as to produce a hole, which is widened by the action of the boring instrument. Presently, the blades of the ovipositor separate, and a single egg is seen to pass between them, so that it is lodged at the bottom of the hole. Into the same aperture is then poured a slight quantity of an irritating fluid, and the insect flies away, having completed her task. The whole proceeding, indeed, is, with the exception of the deposition of the egg, precisely the same as that which takes place when a wasp uses its sting, the ovipositor and sting being but two slightly different forms of the same organ, and the irritating fluid of the cynips being analogous to the poison of the wasp.

The effect of the wound is very remarkable. The irritating fluid which has been projected into the leaf has a singular effect upon its tissues, altering their nature, and developing them into cells filled with fluid. As long as the leaf continues to grow, the gall continues to swell, until it reaches its full size, which is necessarily variable, being dependent on that of the leaf. I have, for example, many specimens of these galls, of different sizes, from which the insects have escaped, showing that they had attained their full size. On the juices of the gall the enclosed insect lives, until it reaches its full term of imprisonment, when it eats its way through the gall and emerges into the world. In some cases, it undergoes the whole of its change within the gall, but in others, it makes its way out while still in the larval state, burrows into the earth, and there changes into the pupal and perfect forms.

To the unassisted eye, the insect which forms the leaf-gall presents no especial attraction, as it is simply, to all appearance, a little black fly. When placed under the microscope, however, it soon proves to be a really beautiful creature, though not possessing the brilliant and gem-like hues which distinguish many of its relatives. The body still retains its blackness, but has a soft tint on account of the white and shining hairs with which it is thickly studded. The eyes are large, stand boldly from the head, and the many lenses of which these organs are composed are so boldly defined, that even in so small an insect they can be distinguished with a very low power of the microscope. Indeed, the inch-and-a-half object glass is quite powerful enough to define them, while the half-inch glass makes them look like the pits in a lady's thimble.

The chief beauty of the insect, however, lies in the wings, which are very large in proportion to the size of the owner, are traversed by a few, but strong nervures, and glow with a changeful radiant lustre, like mother-of-pearl illuminated with living light. In order to see these wings properly, the insect should be laid on some black substance, and the light concentrated upon them by the various means which a microscopist can always employ.

THE oak is a tree that seems to be especially loved by gall-insects, which deposit their eggs in its leaves, its twigs, its flowers, and even in its roots. One of the most familiar examples of oak galls is that which is called the oak-apple, and which is produced by a species of insect called *Cynips terminalis.* Although the insect is not of very great size, the gall which it produces is sometimes enormous, being as large as a common golden pippin or nonpareil apple, and therefore very conspicuous upon the tree. It is coloured in the same manner as the cherry-gall, but seldom so brilliantly, and the exterior is not so smooth and polished.

The resemblance to a veritable fruit is much closer at the beginning of the season than in the autumn, as a number of small leaf-like projections surround its base, just as if they were a half-withered calyx. These, however, fall off as the summer advances, and are no more seen.

If the oak-apple be cut with a knife, the first touch of the steel betrays a marked difference between its substance and that of the cherry-gall. Its texture is neither so firm nor so juicy, but is of a softer, drier, and more woolly character. Moreover. the knife passes through several resisting substances, which, when the gall is quite severed, prove to be separate cells, each containing a grub. From each of these cells, which are extremely variable in number, a kind of fibre runs toward the base of the gall, and it is the opinion of some naturalists that these fibres are in fact the nervures of leaves which would have sprung from the bud in which the gall-fly has deposited her eggs, and which, in consequence of the irritating fluid injected into the tree, are obliged to develop themselves in a new manner.

To procure the insects of this and many other galls is no very difficult task. The branch to which they adhere should be cut off, and placed in a bottle of water, and a piece of very fine gauze tied net-wise over it. The insects, although they can eat their way out of the gall in which they had been bred, never seem to think of subjecting the gauze to the same process, and therefore can be always secured. It is needful, however, to procure galls which are tolerably near their full age, as a branch can only be kept alive for a limited time, and if the supply of nourishment be cut off by the death of the branch, the enclosed insect becomes stunted, if not deformed.

The galls produced by Cynips terminalis are those which are so greatly in request upon the twenty-ninth of May, and which, when covered with gold-leaf, are the standards under which the country boys are in the habit of levying contributions. A figure of this gall is seen in the illustration.

SOME years ago, when I was calling at the office of the *Field* newspaper, then recently started in its race for popularity, I was shown some oak-branches containing a vast number of hard, woody, spherical galls, and asked if I could tell the name of the insect which had produced them. They had recently made their appearance in the country, and no one knew anything about them. A branch beset with these galls is shown in the right hand upper corner of the illustration, the figures being necessarily much reduced.

I was totally unacquainted with them, but, in the following year, found many of them on Shooter's Hill, in Kent, where the growth of oaks is very dense. At the present day they have increased so rapidly that they outnumber almost every species, if we except the tiny spangle-galls; and I have bred great quantities of the insect. The creature which made them is named *Cynips Kollari*, in honour of the celebrated entomologist, and is plentiful on the Continent. I believe that it has long been known in Devonshire, though in Kent it has only recently made its appearance.

The galls produced by this insect are wonderfully spherical, of a brown colour, smooth on the exterior, and about as large as white-heart cherries. Each contains a single insect, which undergoes all its changes within the gall, and eats it way out when it has attained the perfect form. Occasionally two galls become fused together, and in my collection there is a very curious example of these twin galls. They form a figure like that of a rude hour-glass, and each portion has contained an insect. The inhabitant of one portion has eaten its way out and escaped, but the other has met with a singular fate. By some untoward error, it has taken a wrong direction, and instead of issuing into the world in the ordinary way, has hit upon the neck which connects the two galls, so that, instead of merely piercing half the diameter of the gall, it would have been forced to gnaw a passage equal to three half diameters.

Natural powers are always adjusted to the work which their possessors have to perform. The insect was gifted with the capability of eating her way through the walls of her own habitation, but not with the power of making a passage through another gall afterwards. As a natural consequence, she has died from exhaustion before she could emerge into the air; and when I cut the double gall, in order to see how the inmates had fared, I found the dead insect lying near the middle of the second gall, so that she was even farther from the outer air than when she started on her course.

The Cynips Kollari is larger than the generality of the family, equalling a small house-fly in dimensions. Its colour is pale brown. A figure of the insect may be seen in the illustration.

NEARLY in the centre of the illustration is seen a figure of the well-known gall that is so common on the rose, whether wild or cultivated, and which is popularly known by the name of BEDEGUAR. This gall is caused by a very tiny and very brilliantly-coloured insect, named *Cynips rosæ*, which selects the tender twigs of roses, and deposits its eggs upon them.

I have now before me quite a collection of these galls, some of which are so variable in shape that they scarcely seem to have been made by the same species of insect. When the Cynips rosæ deposits her eggs upon the rose, the effects are rather remarkable. Each egg becomes surrounded with its own cell or gall, and the whole of them become fused into one mass. The exterior of these galls is not smooth, like that of the specimens which have been described, but is covered with long, many-branched hairs, which stand out so thickly that they entirely conceal the form of the gall itself.

The number of galls in a single Bedeguar is mostly very great. A specimen of average size, taken at random from the drawer in which the galls are kept, was, when fully clothed, as large as a golden pippin. When the hairy clothing was removed, its size notably diminished, and it was then seen to be composed of a large number of woody tubercles, varying much in size and shape. Their average dimensions, however, are about equal to those of an ordinary pea. The tubercles in question are fused together more or less strongly, some falling off at a slight touch, while others cannot be separated without the use of the knife. There are about thirty-five of these wooden knobs.

On selecting one of the knobs, and examining it, a few very small circular holes are seen, showing that the insects have made their escape from the cells. Indeed, one or two of the insects were found entangled amid the dry and crisp hairs that surrounded the gall, and thus formed a second barrier, which they could not penetrate. When, however, a sharp knife is carefully used, the woody tubercle can be laid open in several directions, and then proves to be a congeries of cells fused together into one mass, and varying from four to twenty in number, according to the size of the insect. Perhaps, on an average, ten cells may be reckoned in each knob.

The cells are of different sizes, some being more than ten times as large as others. The superior dimensions of the cell seem to be obtained at the expense of the walls, so that the large cells can be broken by the finger and thumb, while the small cells cannot be opened without the knife.

The insects themselves are equally variable, some being mere dots of shining blue and green, while others are about as large as the common red ant of the garden, but with plumper bodies. In consequence of these two facts, the large, strong-jawed insect can easily make its way through the comparatively thin walls of the large cell in which it was enclosed, while the small and necessarily weak-jawed specimens are utterly unable to pierce the walls of their cells, which are so thick that they must bore a hole equal in length to that of their whole body before they can escape into the air. Consequently, the great mass of the insects that are found in the cells are the small specimens, the larger having made their escape. I find that on an average twenty small insects are thus found in proportion to one of the larger kind.

THERE is another gall, very common in England, which is found upon the oak, and which is generally thought, by persons who are unacquainted with botany or entomology, to be the bud which naturally grows upon the tree.

In these curious galls, the excrescences with which they are covered take the form of leaves instead of hairs, as is the case with the bedeguar and many other galls. These bud-like objects may be found on the young twigs, and may be easily recognised by their shape, which somewhat resembles that of a pine-apple, and the curious manner in which their leafy covering lies regularly over them, like the tiles upon an ornamental roof. The size of the gall is rather variable, but it is, on an average, about as large as an ordinary hazel-nut.

The gall is so wonderfully bud-like that I have known the two objects to be confounded—the immature acorns in their cups to be carried off as galls, while the real galls were left on the tree. The incipient naturalist who made the mistake kept the buds for some eighteen months, and was sadly disappointed to find that no insects were produced from them.

The insect whose acrid injection produces this curious effect upon the tree is rather larger than the leaf-gall insect, and is of more slender proportions. It has been suggested that the object of the leafy or hairy covering is, that the insect, which remains in the gall throughout the winter, should have a warm house by which it may be protected from the chilling frost as well as from the wind and rain.

IF the reader will again refer to the illustration, he will see that from the same branch on which the *Cynips Kollari* has formed so many galls, depend two slender threads supporting one or two globular objects. These are popularly called CURRANT-GALLS, because they look very much like bunches of currants from which the greater part of the fruit has been removed. Their colour, too, is another reason for giving them this name, as they are sometimes scarlet, resembling red currants, and sometimes pale cream colour, thus imitating the white variety.

These galls are placed upon the catkins of the oak, which are forced to give all their juices to the increase of the gall, instead of employing them on their own development. Some authors think that the insect which forms them is a different species, while others think that the galls are the production of the same insect which forms the leaf-gall, the punctures being made in the stalk of the catkin and not in the nervure of the leaf.

That this supposition may be correct is evident from the fact that the same insect which forms the oak-apples does also deposit its eggs in the root of the same tree, causing large excrescences to spring therefrom, each excrescence being filled with insects. I have often obtained these root-galls, several of which are now before me, some having been cut open, in order to show the numerous cells with which they are filled, and others left untouched, in order to exhibit the form of the exterior. Being nourished by the juices of the root, they partake of the sombre hues which characterise the part of the tree from which they spring, and do not display any of the colours which are seen on the oak-apples which spring from the twigs.

There are, however, distinct species of gall insects which pierce the roots of the oak-tree. One of them is termed *Cynips aptera*, and makes a pear-shaped gall about one-third of an inch in diameter. Each gall contains a single insect, and a number of the galls are often found attached by their narrow end to the root-twigs of the tree, something like a bunch of nuts on a branch. There is another insect which is termed *Cynips quercus-radicis*, which forms a many-chambered gall of enormous size, containing a small army of insects. Mr. Westwood mentions that one of these galls in his possession was five inches long, one inch and a quarter wide, and produced eleven hundred insects, so that the entire number was probably fourteen or fifteen hundred.

No one who is accustomed to notice the objects which immediately surround him can have failed to observe the curious little galls which stud the leaves of several trees, and which are appropriately called SPANGLE-GALLS, because they are as circular, and nearly as flat, as metallic spangles.

These objects had been observed for many years, but no one knew precisely whether their growth was due to animal or vegetable agency. That their substance was vegetable was a fact easily settled, but some botanists thought that they were merely a kind of fungus or lichen, while others supposed that they were the work of some parasitic insect.

When closely examined, these 'spangles' are seen to be discs, very nearly but not quite flat, fastened to the leaf by a very small and short central footstalk. Réaumur set at rest the question of their origin by discovering beneath each of them the larva of some minute insect, but he could not ascertain the insect into which the larva would in process of time be developed. The task of rearing the perfect insect from the gall is exceedingly difficult, the minuteness of the species and the peculiar manner in which the development takes place, being two obstacles which require a vast expenditure of care and patience before they can be overcome.

Supposing a branch containing a number of infested leaves to be placed in water and surrounded with gauze, it will die in a week or two, and yet there will be no sign of an insect. If the branch be kept until the winter has fully set in, the desired insects will still be absent, and the experimenter will probably think that his trouble has been thrown away. The real fact is, that the little insects are not developed until the spring of the following year, and that they pass through their stages of the pupal and perfect forms after the leaves have fallen, and while they are still lying on the ground.

Mr. F. Smith, who has given so much time and research to the history of the hymenoptera, has discovered the insect that inhabited the galls to be *Cynips longipennis*, and has remarked that the perfect insects do not make their appearance until the month of March. I have had many specimens of this tiny and beautiful insect.

CHAPTER XXVI.

PARASITIC NESTS—(concluded).

The Oak-tree, and its aptitude for nourishing Galls—COMPOUND GALLS, or one Gall within another—The SENSITIVE GALL of Carolina—Galls and the Insects which caused them—Colours of Galls—Whence derived—The Galls of various trees and plants—The Cynips parasites upon an insect—Galls produced by other insects—Mr. Rennie's account of the BEETLE GALL of the Hawthorn—The BEETLE GALL of the Thistle—DIPTEROUS GALL-MAKERS—Animal Galls—The CHIGOE and its habits—Its curious egg-sac—Difficulty of extirpating it—The penalty of negligence—The BREEZE FLIES and their habitations—WURBLES and their origin—Their influence upon cattle—The CLERUS and its ravages among the hives—The DRILUS, its remarkable form and the difference between the sexes — The curious habitation which it makes.

THE reader cannot but notice the singular aptitude possessed by the oak-tree for nourishing galls. No part of the tree seems to escape the presence of a gall of some sort, diverting its vital powers into other channels. The tree, however, does not appear to suffer from them, and it is just possible that they may be useful to it. The leaves are studded with galls, and so are their stems. The branches are covered with galls of various shapes, sizes, and colours, some bright, smooth, and softly coloured, like ripe fruit, others hard, harsh, spiny, and rough, as if the very essence of the gnarled branches had been concentrated in them. There are galls upon the flowers, galls upon the trunk, and even galls upon the root.

Some oak-galls may be called compound galls. M. Bosc mentions a small gall which is found upon the American oak. It is not larger than a pea, and if shaken is found to contain some hard substance loosely lodged in its interior. When the gall is cut open, a very curious state of things is seen. The walls are very thin, so that in spite of the small dimensions, the cell is larger than that of many cynipidæ. Within the cell, no insect is discovered, but in its place a little spherical object, about as large as a No. 5 shot, which is very hard, and rolls about freely in the interior. If this be opened, the larva is found within it, reminding the adept in fairy lore of the white cat whose gifts were enclosed in a succession of nuts, each within the other. How these singular little cellules are made is not known, though their discoverer expended great trouble and patience upon them.

The same naturalist mentions another species of gall, also found upon the oak in Carolina. It is spherical, covered with prickles like a thistle, and beset with a thick downy covering of rather long hair. Many other galls possess these characteristics, but the most curious point connected with this species is, that the hairs are as mobile as those of the sensitive plant, and as soon as they are touched, sink down, and never afterwards regain their former position.

THE size of a gall is no criterion of the dimensions or numbers of the insect which made it. Even in the galls which infest the oak, the smallest galls often furnish the largest insects, and in some specimens brought from Greece, the gall is as large as an ordinary black-currant, while the cell would contain a red-currant, showing that the inhabitant of the cell must be a large one in order to fill it. Again, although the oak-apple and rose-bedeguar do contain a great number of insects, there are many examples where galls scarcely so large as a pea contain from ten to fifteen insects, while the ink-gall and the large Hungarian gall are inhabited by a single insect.

One of the most curious problems is, to my mind, that of the brilliant colours with which many of these galls are decorated. That the rose-bedeguar should be so beautifully adorned with scarlet and green is a fact which does not seem to excite any astonishment, inasmuch as it may be said that the colours which ought to have been developed in the petals and the leaves have been diverted from their proper course, and forced to exhibit themselves in the gall.

Botanists and physiologists will see that this idea is quite groundless, but to the uninstructed and popular mind it has a sort of plausibility that often commands assent. But when we come to the oak-tree the case is at once altered, and some other cause must be found for the lovely colours of its galls. The cherry-galls are as brightly coloured as any apple, and the soft hues of the oak-apple are nearly as beautiful though not so brilliant. Yet the oak possesses no such store-house of colour as is popularly attributed to the rose. Its leaves are simple green, and its flowerets are so colourless as scarcely to be distinguished by the unassisted eye.

Whence then are derived these beautiful colours? Some hasty observers, who have neglected the first rule of logic, and drawn an universal conclusion from particular premises, have said that the colours of the gall are derived from the insect; adducing, as a proof of their assertion, the brilliant colours which equally deck the rose-bedeguar and the *Cynips rosæ* from which it sprang. But if they had only followed the example of careful naturalists, who, like Dr. Hammerschmidt, have examined and drawn between two and three hundred species of galls, so hasty a generalisation would never have been made. The cherry or leaf-gall of the oak is every whit as gorgeously coloured as the bedeguar of the rose, while the insect that made it is quite black. It is true that the diaphanous wings glitter as if they were made of polished gems; but this appearance is due, not to the wings themselves, but to the myriad hairs with which they are regularly studded, each hair acting as a miniature prism by which the light is refracted and broken into the resplendent hues of the rainbow.

MANY other trees besides the oak are chosen by certain species of gall-fly, and even the herbs and flowers do not escape the ravages of these remarkable insects. The white poppy, from which is obtained the opium of commerce, is attacked by a species of gall-fly, which lays its eggs in the large head, or pod, and sometimes does much damage to the plant, the delicate divisions between the seed-vessels being rendered quite hard and solid, and the pod itself deformed. Mr. Westwood has described a species of gall-fly which infests the turnips, and another species is known to lay its eggs upon wheat.

As if to show that the family of Cynipidæ is really related to the ichneumons, it has been discovered that some species of this family are actually parasitic upon other insects. In treating of this remarkable fact, Mr. Westwood writes as follows:—'The relations of these insects with the following families (*i.e.* Evanidæ and Ichneumonidæ) have been already noticed. It had always appeared to me contrary to nature that a tribe of vegetable-feeding insects should be arranged in the midst of parasites; nor was it until I had an opportunity of ascertaining the parasitic habits of some of the species of the family, that I was enabled to form a just notion as to the true value of the parasitic or herbivorous nature of these insects. In June, 1833, I detected a minute species, *Allotria victrix*, in the act of ovipositing in the body of a rose-aphis, and I subsequently succeeded in hatching specimens of the perfect insect from infested aphides.'

A figure of the tiny insect is given, as it appeared while in the act of depositing its eggs, and has a rather remarkable effect from the fact that the very minute dimensions of the parasite make the aphis look quite a large insect. Other species of this family are also known to be parasitic. The rose-aphis is certainly infested by two species of gall-fly, and probably by more, while the aphides which are found on the willow, the cow-parsnip, and other plants, also fall victims to the Cynipidæ. There is one genus of this family, called *Figites*, which is parasitic on the larva or pupa of certain dipterous insects.

THE Cynipidæ are not the only insects that produce galls upon different plants. For example, several species of beetle are known to pass their earlier stages in swellings produced by the puncture of the parent insect. There is a little weevil of a greyish brown, which is mentioned by Mr. Rennie, as forming a gall upon the hawthorn.

'In May, 1829, we found on a hawthorn at Lee, in Kent, the

leaves at the extremity of a bunch neatly folded up in a bundle, but not quite so closely as is usual in the case of leaf-rolling caterpillars. On opening them up, there was no caterpillar to be seen, the centre being occupied with a roundish, brown-coloured, woody substance, similar to some excrescences made by gall insects (*Cynips*).

'Had we been aware of its real nature, we should have put it immediately under a glass, or in a box, till the contained insect had developed itself; but instead of this, we opened the ball, where we found a small yellow grub coiled up, and feeding on the exuding juices of the tree. As we could not replace the grub in its cell, part of the wall of which we had unfortunately broken, we put it in a small pasteboard-box with a fresh shoot of hawthorn, expecting that it might construct a fresh cell. This, however, it was probably incompetent to perform; it did not, at least, make the attempt, and neither did it seem to feed on the fresh branch, keeping in preference to the ruins of its former cell.

'To our great surprise, although it was thus exposed to the air, and deprived of a considerable portion of its nourishment, both from the fact of the cell having been broken off, and from the juices of the branch having been dried up, the insect went through its regular changes, and appeared in the form of a small greyish brown beetle of the weevil family.

'The most remarkable circumstance in the case in question, was the apparent inability of the grub to construct a fresh cell after the first was injured,—proving, we think, beyond a doubt, that it is the puncture made by the parent insect when the egg is deposited that causes the exudation and subsequent concretion of the juices forming the gall.' Although the insect in question succeeded in attaining the perfect state, it would probably be of stunted growth in consequence of the deprivation of food. Such, at all events, is the case with insects of other orders, when their supply of food is at all checked while they are in the larval state.

THERE is another weevil, scientifically called *Cleonus sulci-*

rostris, which is one of the gall-makers. It is one of the largest of the British weevils, being more than half an inch in length, and is very simply clad in grey and black.

If the reader desires to discover the larva of the beetle he may probably be successful by going to any waste spot where thistles are allowed to grow, and examining them carefully about the stems and roots. Nothing is more common than to find the stems of thistles swollen in parts, and in many cases the root is affected as well as the stem. Fortunately for the gardener, who hates thistles, even though he should be a Scotchman, as is so often the case with skilled gardeners, the larva of the Cleonus feeds on the juices of the plant at the expense of its life, so that the thistle dies just before the seed is developed, and a further extension of the plant is thereby prevented.

THERE are also gall-making insects among the Diptera. Such, for example, is the THISTLE-GALL FLY (*Urophora Cardui*), which produces large and hard woody galls upon the thistle, as well as several species of the larger genus Tephritis, some species of which live in the parts of fructification of several flowers, the common dandelion being infested by them.

WE must now glance at a few of the insects that are parasitic upon other animals. Their numbers are very great, but we must restrict ourselves to those which construct some sort of a habitation.

The only insect which can be said to be parasitic on man, and at the same time to form a habitation, is the celebrated CHIGOE (*Pulex penetrans*), otherwise called the JIGGER, or EARTH FLY. This terrible pest is a native of Southern America and the West Indian islands, and is too well known, especially by the negroes and natives.

This insect, which is closely allied to the common flea, and much resembles it in general appearance, contrives to hide itself under the nails of the fingers or toes, usually the latter. Having gained this point of vantage, it proceeds very gradually to make its way under the skin, and, strange to say, does so without causing any pain. There is a slight irritation, rather pleasing than otherwise, to which a novice pays no attention, but which puts an experienced person on his guard at once.

The male Chigoe is innocent of causing any direct injury to man, the female being the cause of all the mischief. As soon as she is settled, her abdomen begins to swell until it becomes quite globular, and of great comparative size, and containing a vast quantity of tiny eggs. Pain is now felt by the victim, who generally has recourse to the skilful old dames, who have a kind of monopoly of extracting Chigoe 'nests.' With a needle, they carefully work round the globular body of the buried insect, taking great care not to break it, as if a single egg remains in the wound, all the trouble is wasted. By degrees they gently eject the intruder, and exhibit the unbroken sac of eggs with great glee. To prevent accidents, however, the wound is filled with a little Scotch snuff, which certainly causes rather a sharp smarting sensation, but effectually destroys any egg or young insect that may perchance have escaped notice.

Europeans and natives of the better caste escape easily enough, because they always take warning by the first intimation of a Chigoe's attack, and generally succeed in killing her before she has succeeded in burying herself. Moreover, the shoes and stockings of civilised man protect his feet, and the gloves guard his hands, so that the insect does not find many opportunities of attacking the white man.

But the negroes, and especially the children, suffer terribly from the Chigoe. Children never are very apt at sacrificing the present to the future, and the negro child is perhaps in this particular the least apt of all humanity. The Chigoe is in consequence seldom disturbed until it has made good its entrance, and even then would not be mentioned by the child, on account of the pain which he knows is in store for him. But the experienced eyes of the matrons are constantly directed to the feet of their children, and if one of them is seen to hold his toes off the ground as he walks, he is immediately captured and carried off to the operator, uttering dismal yells of apprehension.

He certainly has good reason for his fears. The Chigoe nest

is duly removed, and then, partly to prevent the hatching of any egg that may have escaped during the operation, and partly to punish the delinquent for his disobedience, the hollow is filled, not with snuff (which is too valuable a substance to be wasted), but with pounded capsicum. The discipline is certainly severe, but it is necessary. After a child has once paid the penalty of negligence, he seldom chooses to bring such a punishment on himself a second time, and as soon as he feels the first movements of a Chigoe, away he goes to have it removed before it can burrow under the skin.

If the Chigoe be allowed to remain, the results are disastrous. Swellings make their appearance along the limbs, the glands become affected, and if the cause is permitted to remain undisturbed, mortification takes place, and the sufferer dies. So the red-pepper discipline, severe as it may be, is an absolute necessity with those who are unable to reason rightly, or to exercise forethought for the future. Every evening the negro quarter of the villages is rendered inharmonious by the outcries of the children who have neglected to report themselves in proper time, and who in consequence are suffering the penalty of their negligence.

THERE are some insects which produce upon animals certain swellings which are analogous to the galls upon trees. Such, for example, is the well-known BREEZE FLY (*Œstrus bovis*), which is so troublesome to cattle. The larvæ of this insect live under the skin of the animal, and in some manner raise a large swelling, that is always filled with a secretion on which they live. In fact, the swelling is a gall produced on an animal instead of a plant, and the enclosed insect feeds in a similar manner upon the abnormal secretion which is induced by the irritation of its presence.

The larvæ are fat, soft, oval-bodied creatures, and are notable for the flattened end of the tail, on which are placed two large spiracles or breathing-holes.

Although the larva which inhabits the vegetable gall seems to have but small need of air, and to all appearance can exist without any apparent channel of communication with the external atmosphere, such is not the case with the inhabitant of the animal gall. An opening is always preserved in the upper part of the swelling, and the tail of the grub is tightly pressed against the aperture so as to ensure a constant supply of air.

In the months of May and June, these swellings may be found in great plenty. They are mostly seen upon young cattle, and as a general rule are situated close to the spine. So common indeed are they, that out of a whole farm-stock of cattle I have seen almost every cow under the age of four years attacked by the Breeze Fly, and counted from two or three to twelve or fourteen upon a single animal. It is said that as many as forty have been detected upon a single cow, but such an event has not come within my own observation.

The swellings caused by the Breeze Fly are called Wurbles, or Wornils, and can be easily detected by passing the hand along the back. Strangely enough, the cow does not appear to feel any pain from the presence of these large parasites, nor does she suffer in condition from them, although it would seem that they must keep up a continual drain upon the system. Indeed, some experienced persons have thought that, instead of being injurious, they are absolutely beneficial.

When the grub has reached its full development, it pushes itself backwards out of the gall, and falls to the ground, into which it burrows. Presently, the skin of the pupa becomes separated from that of the larva, and the latter dies, and becomes the habitation in which the pupa lives. The head portion of the skin is so formed that it flattens when dry, and can easily be pushed off, like the lid of a box, permitting the perfect fly to escape. Even when the insect is still in its pupal condition this lid can be removed, so that the pupa can be seen within its curious habitation. I may mention here that insects which are thus covered while in their pupal state, so as to show no traces of the creature within, are said to undergo a 'coarctate' metamorphosis. Nearly all the diptera are examples of the coarctate insects.

BEFORE we close the subject of parasites, it will be needful to give a brief account of one or two parasitic insects which possess points of peculiar interest in the habitations which they make, or in the places wherein they find their abode.

One of these insects is rather a pretty beetle, termed *Clerus alvearius*. In its perfect state it is innocent enough, but in its larval state it is so destructive among the hives, that all bee-keepers will do well to destroy every Clerus that they can catch. It is generally to be found on flowers, licking up their sweet juices by means of a brush-like apparatus attached to the mouth. The wing-cases of most of the species are bright red, barred or spotted with purple.

The larva is of a beautiful red, and is hatched from an egg placed in the cell occupied by the bee-grub. As soon as it is hatched, it proceeds to feed upon the bee-grub, and devours it. Unlike many insects with similar habits, it is not content with a single grub, but proceeds from cell to cell, devouring all their inhabitants. When it has eaten to the full, it conceals itself in the cell, and spins a cocoon of rather small dimensions in comparison with its own size. In process of time, it is developed into a perfect insect, and then breaks out of its cocoon and leaves the hive, secure from the bees, whose stings cannot penetrate the horny mail in which it is encased.

THERE is another beetle which is parasitic upon snails, and which, in its larval and pupal states, is only to be found within those molluscs. Its scientific name is *Drilus flavescens*, the latter name being given to it in honour of its yellow-tinted wing-cases, which present a pretty contrast with the black thorax. It is a little beetle, scarcely exceeding a quarter of an inch in length, and is remarkable for the beautiful comb-like antennæ of the male. As for the female, she is so unlike her mate that she has been described as a different insect. She has no pretensions to beauty, and can scarcely be recognised as a beetle, her form being that of a mere soft-bodied grub. Moreover, the size of the two sexes is notably different. The male is, as has already been observed, only about a quarter of an inch long, while the female is not far from an inch in length, and is broader than the length of her mate, antennæ included.

This curious insect lives in the body of snails, the common banded snail of our gardens being its usual prey. When it is about to change into the perfect state, it makes a curious cocoon, of a fibrous substance, which has been well likened to common tobacco, the scent as well as the form increasing the resemblance. The grub or larva of this beetle bears a very great resemblance to the perfect female, and indeed is so similar that none but an entomologist could distinguish the two creatures. It is furnished with a number of false legs, as well as with a forked appendage at the end of the tail, by which it is enabled to force its way into the body of its victims. The head is pointed, and the jaws are very powerful.

CHAPTER XXVII.

BRANCH-BUILDING MAMMALIA.

The DORMOUSE in Confinement, and at Liberty—Nest of the Dormouse—Its position, materials, and dimensions—Entrance to the nest—The winter treasury—The LOIRE and the LEROT.

WE now come to another division of the subject, namely, the nests that are built in branches; and adhering to the system which has been followed through the progress of the work, we shall take first the branch-building mammalia.

There are but few mammals which can be reckoned in this division, but our little island produces two of them, namely, the squirrel and the DORMOUSE (*Muscardinus avellanarius*). The former of these animals have been already described at page 118.

The pretty little brown-coated, white-bellied Dormouse is familiar to all who have been fond of keeping pets. There is no difficulty in preserving the animal in health, and, therefore, it is a favourite among those who like to keep animals and do not like the trouble of looking after them. It is, however, rather an uninteresting animal when kept in a cage, as it sleeps during the greater part of the day, and the sight of a round ball of brown fur is not particularly amusing.

When kept in confinement, it is obliged to make for itself a very inartificial nest, because it is deprived of proper materials and a suitable locality. It does its best with the soft hay and cotton wool which are usually provided for it, but it cannot do much with such materials. But when in a state of liberty, and able to work in its own manner, it is an admirable nest-maker. As it passes the day in sleep, it must needs have some retired domicile in which it can be hidden from the many enemies which might attack a sleeping animal.

One of these nests forms a part of my collection. It was situated in a hedge about four feet from the ground, and was placed in the forking of a hazel branch, the smaller twigs of which form a kind of palisade round it. The substances of which it is composed are of two kinds, namely, grass-blades and leaves of trees, the former being the chief material. It is exactly six inches in length by three inches in width, and is constructed in a very ingenious manner, reminding the observer of the pensile nests made by the weaver birds.

Two or three kinds of grass are used, the greater part being the well-known sword-grass, whose sharp edges cut the fingers of a careless handler. The blades are twisted round the twigs and through the interstices, until they form a hollow nest, rather oval in shape. Towards the bottom the finer sorts of grass are used, as well as some stems of delicate climbing weeds, which are no larger than ordinary thread, and which serve to bind the mass together. Interwoven with the grass are several leaves, none of which belong to the branch, and which are indeed of two kinds, namely, hazel and maple, and have evidently been picked up from the ditch which bounded the hedge. Their probable use is to shield the inmate from the wind, which would penetrate through the interstices of the loosely woven grass-blades.

The entrance to the nest is so ingeniously concealed, that to find it is not a very easy matter, even when its precise position is known, and in order to enter the nest, the Dormouse is obliged to draw aside certain broad grass-blades which are ingeniously disposed over the entrance so as to hide it. The pendent pieces of grass that are being held aside by the little paw are so fixed, that when released from pressure, they spring back over the aperture and conceal it in a very effectual manner.

Although the Dormouse uses this aerial house as a residence, it does not make use of it as a treasury. Like many other hibernating animals, it collects a store of winter food, which generally consists of nuts, grain, and similar substances. These treasures are carefully hidden away in the vicinity of the nest.

During the winter the animal does not feed much upon its stores, inasmuch as it is buried in the curious state of hibernation during the cold months. At the beginning of spring, however, the hibernation passes off, and is replaced by ordinary sleep, with intervals of wakefulness.

Now, while the animal hibernates, the tissues of the body undergo scarcely any change, even though no nutriment be taken. But, as soon as the creature resumes its ordinary life, waste goes on, and the creature soon feels the pangs of hunger. As the food of the Dormouse consists chiefly of seeds and fruits, it could not find enough nourishment to support the body, and would therefore perish of hunger but for the stores which instinct had taught it to gather in the preceding autumn.

CHAPTER XXVIII.

FEATHERED BRANCH-BUILDERS.

The ROOK and its nesting-place—Materials and structure of the nest—Some habits of the Rook—The CROW—Difference between the nest of the Rook and the Crow—The HERON and its mode of nidification—The Heronry at Walton Hall—Rustic ideas respecting the Heron's nest—The CHAFFINCH—Locality and structure of its beautiful nest—Mode of obtaining materials—The GOLDFINCH and its home—Distinction between the nests of the Goldfinch and Chaffinch—The BULLFINCH—Locality and form of its nest—Variability of Structure—The GOLDEN ORIOLE and its beautiful nest—Mode of catching the Bird—The YELLOW-BREASTED CHAT and its odd ways—Its courage and affection for its nest and young—Structure of its nest—The RINGDOVE and its curious nest—The MOCKING BIRD—The WATER-HEN and its nesting—Its habit of covering the eggs.

WE pass now to the many birds which build their nests on branches of trees or shrubs, and which may therefrom be termed AERIAL BUILDERS. A vast proportion of the feathered tribes select branches as a site for their habitation, so that only the remarkable examples will be mentioned.

PERHAPS the most conspicuous of all ordinary branch-nests are those which are made by the Rooks and the Crows.

Every one has seen the nests of the former of these two birds. They are large, dark, and are placed upon the topmost boughs of the tree, so that they can be seen at a considerable distance. Their position is evidently intended as a safeguard against the attacks of various enemies, among which the bird-nesting boy is pre-eminently the most dangerous. Scarcely would the boughs endure the weight of a cat or monkey, and so slender are they in many cases, that the spectator wonders how they can support the nest with its living contents of a parent and three or four young.

The foundation of the nest is composed of sticks of various sizes and lengths, all, however, being tolerably light and dry, the Rook generally carrying up the dead branches that have been blown down by the winds of the preceding winter. These are usually interlaced among the spreading branches of a convenient spray, and thus form a rude basket-work, in which will lie the softer materials on which the eggs and young are to repose. The lining is composed almost entirely of long and delicate fibrous roots, which are intertwined, so as to make an interior basket very similar in general construction to the twig basket of the exterior, and being so independent of it that, with a little care, it can be lifted out entire.

On this soft bed are laid the eggs, which are four or five in number, and are rather variable in colour, the usual tint being greenish grey, largely spotted, mottled, and splashed with dark brown, in which a shade of green is visible. They vary in size as well as in hue, and from the same nest I have taken eggs of so different an aspect that a casual observer would probably think them to be the production of distinct birds.

The principal labours of nest-building fall on the young birds, inasmuch as the elders mostly return to the same domicile every successive season, and are seldom obliged to make an entirely new nest. The young builders are sometimes aggrieved at this distribution of labour, and try to equalize it by helping themselves to the sticks belonging to other proprietors. The general community, however, never suffer theft to be perpetrated, and are sure in such a case to scatter the ill-gotten materials, and force the dishonest birds to begin their labours anew.

When the young are launched upon the world and able to get their own living, the nest is used no more, but is abandoned both by parents and young, not to be again used until repaired in the spring of the following year. It is a curious point in the economy of the Rook, that, when it has abandoned its temporary home, it does not choose to repose among the trees on which the nest was made. Mr. Waterton, who possesses invaluable opportunities for studying the habits of this bird, and has developed them to the utmost, makes the following remarks upon the roosting of this bird :—

'There is no wild bird in England so completely gregarious as the Rook, or so regular in its daily movements. The ring-doves will assemble in countless multitudes, the finches will unite in vast assemblies, and waterfowl will flock in thousands to the protected lakes, during the weary months of winter; but when the returning sun spreads joy and consolation over the face of nature, these congregated numbers are dissolved, and the individuals retire in pairs to propagate their respective species. The Rook, however, remains in society the year throughout. In flocks it builds its nest, in flocks it seeks for food, and in flocks it retires to roost.

'About two miles to the eastward of this place are the woods of Nostell Priory, where from time immemorial the Rooks have retired to pass the night. I suspect, by the observations which I have been able to make on the morning and evening transit of these birds, that there is not another roosting-place for at least thirty miles to the westward of Nostell Priory. Every morning, from within a few days of the autumnal to about a week before the vernal equinox, the Rooks, in congregated thousands upon thousands, fly over the valley in a westerly direction, and return in undiminished numbers to the nest, an hour or so before the night sets in.

'In their morning passage, some stop here; others in other favourite places, farther and farther on; some repairing to the trees for pastime, some resorting to the fields for food, till the declining sun warns those which have gone farthest that it is time they should return. They rise in a mass, receiving additions to their numbers from every intervening place, till they reach this neighbourhood in an amazing flock. Sometimes they pass on without stopping, and are joined by those which have spent the day here. At other times they make my park their place of rendezvous, and cover the ground in vast profusion, or perch upon the surrounding trees. After tarrying here for a certain time, every Rook takes wing. They linger in the air for awhile, in slow revolving circles, and then they all proceed to Nostell Priory, which is their last resting-place for the night.

'In their morning and evening passage, the loftiness or lowliness of their flight seems to be regulated by the state of the weather. When it blows a hard gale of wind, they descend the valley with astonishing rapidity, and just skim over the tops of the intervening hills, a few feet above the trees: but when the sky is calm and clear, they pass through the heavens at a great height, in regular and easy flight.'

This custom of the Rooks is the more curious because it is hardly possible to conceive any roosting-place which would be more acceptable to a sensible bird than the woods within the confines of Walton Hall. As has already been mentioned, the birds will occasionally rest for a while in those pleasant woods, though they ultimately take wing for the accustomed roosting-place. There is plenty of space for them; they have their choice of trees on which to settle, and the lofty wall which surrounds them ensures their freedom from all disturbance.

VERY similar in general aspect to the rook, the CROW (*Corvus corone*) builds a nest which resembles that of the rook in outward form, but is easily distinguished by an experienced eye. The lining of the nest is made of animal instead of vegetable substances, hair and wool taking the place of fibrous roots.

Viewed from the foot of the tree, the nest of the Crow is nothing but a large and nearly shapeless bundle of sticks, but when the enterprising naturalist has climbed to the summit of the tree in which it is placed, and can look into the nest, he is always gratified by the peculiarly neat and smooth workmanship of the aerial home. The outside of the nest is rough and rugged enough, but the inner nest, which is made of rabbit's fur, wool, and hair, is woven into a basin-like form, beautifully smooth, soft, and elastic. On this bed repose the eggs, which are somewhat like those of the rook, but darker and greener, and more thickly spotted, though they are extremely variable in size and colour, and sometimes resemble so closely those of the rook that the distinction can hardly be detected.

The Crow always builds at the tops of trees, and has a wonderful knack of choosing those which are most difficult of ascent. The nests are plentiful enough, but the proportion of eggs taken is very small in comparison. There are some nests which baffle almost anyone to rob successfully. An experienced nest-hunter is always endowed with a strong head, and ought to be perfectly at his ease on the summit of the loftiest trees, even though he should be obliged to crawl in fly-fashion under a branch, to hang by one hand while he takes the eggs with the other, or to suspend himself by his legs in order to get at a nest below him. That a nest should escape a properly qualified hunter is simply impossible, but to secure the eggs is quite another matter.

In many cases the nest of the Crow is placed on branches so long and so slender that they will not endure the weight of a small boy, much less of a man, and the only method of getting at it is by bending down the branches. But, when the branches are bent, the nest is tilted over, and out fall the eggs, so that the disappointed hunter loses all his time and trouble.

Possibly this extreme caution may be the result of sad experience, for, although the generality of Crows' nests are placed in the most inaccessible positions, I have seen and taken many which were so easy of attainment that in a very few minutes I had ascended the tree and returned with the eggs. There are generally four or five eggs, although in some exceptional cases six eggs are said to be laid in a single nest. I never saw more than five, though I have examined very many nests. High as the nest of a Crow may be, it is worthy of an ascent, for, even should it be an old nest and deserted by the original inhabitant, there is always a possibility that it may have been usurped by some hawk, whose beautiful eggs are always considered as prizes.

THERE is a splendid British bird, which is becoming scarcer almost yearly, which makes a nest something like that of the crow and rook, but much larger. This is the HERON (*Ardea cinerea*), one of the very few large birds which still linger among us.

On account of its own great size, the Heron makes a very large and very conspicuous nest, built chiefly of sticks and twigs, and placed on the summit of a tree.

Like the rook, the Heron is gregarious in its nesting, so that a solitary Heron's nest is very seldom seen, though now and then an exception to the general rule is discovered. To watch the manners and customs of this bird is not a very easy task, because the number of heronries in England is very small, and the shy nature of the birds renders them difficult of approach. At Walton Hall, however, the Herons are so fearless, through long-continued impunity, that they will allow themselves to be watched closely, provided that the observer is quiet, and does not make a noise, or alarm the birds by abrupt movements.

It is a very pretty sight to watch the great birds as they go to and from their nests, bringing food to their young, or flying to the lake in search of more fish. Numbers of the Heron may be seen at the water's edge, sometimes standing on one foot, with their long necks completely hidden, and their bayonet-like beaks projecting from their shoulders. For hours the birds will retain this attitude, which to a human being would be the essence of discomfort, and it is really wonderful how they can keep up for so long a time the muscular energy which is expended in holding up the spare leg and keeping it tucked under the body.

Now and then, one of the Herons seems to wake up, and after a stretch of the neck and a flap of the wings, walks statelily and deliberately into the water, through which it stalks, examining every inch of bank and every cluster of weeds as it passes along. Presently the bird pauses, and remains quite still for some time, when the long neck is suddenly darted forwards, the beak disappears for a second among the reeds, and then emerges, with a fish, frog, or water-rat in its gripe. The real beauty of the Heron can never be appreciated until it is seen at liberty, and in the enjoyment of its natural life. It suits the locality so well that, when it flies away, the spot has lost somewhat of its charms. As it stands in the water, intent upon catching prey, the drooping feathers of its breast wave gracefully in the breeze, and the ripples of the sunlit water are reflected in mimic waves upon its grey plumed wings.

Generally it cares little for exerting itself until towards the evening, but then it becomes impatient and restless, and is not quieted until it has obtained some food.

Some anglers have an idea that the Heron is one of the birds that ought to be ranked as 'vermin,' thinking that it destroys so many fish, that it ruins an angler's sport. Consequently, they kill the bird whenever they can manage to do so, and flatter themselves that they are doing good service in preserving the breed of fish. Now, even were the entire diet of the Heron to consist of fish, the bird would really do but little harm, because it can only take food in shallow water, and is seldom to be seen more than a yard or two from the bank. But the diet of the Heron is by no means exclusively of a fishy nature, inasmuch as the bird eats plenty of frogs and newts, and will often secure a water-rat even when fully grown. It is seldom that fish which are of any value to the angler come into water in which the Heron could catch them, and even if they did so, their size would prevent the bird from taking them.

At Walton Hall, where the Herons breed largely, and where they procure nearly all the food for themselves and young out of the lake, there is no lack of fish, as may be practically proved by anyone who is permitted to cast a line into the water. I am a very poor fisherman, and yet I never found any difficulty in taking in the course of the morning quite as many fish as could easily be carried home.

So far indeed is the Heron from injuring the interests of the angler, that it is a positive benefactor. Mr. Waterton, who was obliged by the continual burrowing of water-rats to drain and fill up a series of large ponds, makes the following remarks on the bird :—'Had I known then as much as I do now of the valuable services of the Heron, and had there been a good heronry near the place, I should not have made the change. The draining of the ponds did not seem to lessen the number of rats in the brook; but soon after the Herons had settled here to breed, the rats became exceedingly scarce, and now I rarely see one in the place where formerly I could observe numbers sitting on the stones at the mouth of their holes, as soon as the sun had gone below the horizon.'

When the Heron flies to its nest from any great distance, it generally ascends to a considerable height, and is in the habit of uttering a curious and very harsh cry, which at once tells the naturalist that a Heron is on the wing. When a Heron passes immediately over the observer, the effect is very remarkable, the long, stretched-out legs and neck and slender body looking like a large knitting-needle supported on enormous wings.

To see the Heron alight on its nest or on a branch is rather a curious sight. The bird descends, drops its long legs, places its feet on the branch, and then flaps its huge wings as if to get its balance before it settles down. The rustics have an idea that a Heron is obliged to allow its legs to dangle on either side of the nest while it sits on its eggs, and some will aver that a hole is made in the nest through which the legs can be thrust. It is scarcely necessary to say that the construction of a bird's legs prevents it from assuming such an attitude, and that the long Heron can sit as easily upon its pale green eggs as the short-limbed domestic fowl on her white eggs.

Some of our common British birds build nests that can vie, in point of beauty and delicacy, with any nest made by birds of other lands. It is scarcely possible to conceive a nest which is more worthy of admiration than that of the Long-tailed Titmouse, which has already been described; and in their own way, the houses erected by the Chaffinch and Goldfinch are quite as beautiful. As there are some points of similarity in the two nests, they will be mentioned in connection with each other.

First, we will take the nest of the Chaffinch (*Fringilla cœlebs*).

Although the beautifully-spotted eggs are plentiful in the collection of every nest-hunting schoolboy, they do not come into his little museum for some time. The eggs of the blackbird, thrush, and hedge-warbler are generally the first to be found, because the nests in which they are contained are so conspicuous. But the nest of the chaffinch is never easily seen, and its discovery requires a special training of the eye.

An experienced nest-hunter will always detect it, and it is amusing to watch the bewildered expression of a novice to whom a Chaffinch-nest is pointed out, and who cannot see it in spite of all the indications of his instructor. The bird likes to find the fork of a tree or bush, where several branches are thrown out from one spot, and so as to form a kind of cup in which the nest can lie. Tall hawthorns, or even sloe or crab-trees, especially if they grow in thick hedges, are favourite trees with the Chaffinch, and a luxuriant and untrimmed hedgerow is always prolific in Chaffinch-nests.

Within the forked branches, the bird constructs its nest, and does so in rather a singular manner. The chief material is wool, which is matted together so as to form a kind of loose felt, and with this felt are woven delicate mosses, spider-webs, cottony down, and lichens. The last-mentioned materials are stuck most ingeniously upon the outside of the nest, and have the effect of making it look exactly like a natural excrescence from the tree in which it is placed.

This pretty nest is generally deep in proportion to its width, and is lined with hairs, arranged in a most methodical manner, so as to form a cup for the eggs. The hair of the cow is much used by the Chaffinch, which may be seen collecting its stock of hairs from the fields wherein cows are pastured, not plucking them directly from the body of the animal, but searching for them in the crevices of the trees and posts against which the cattle are accustomed to rub themselves. Mostly, the bird can only procure single hairs; but when it is fortunate enough to find a tuft or bunch of hairs, it pulls them out, and works them separately into the nest, so as to ensure the needful uniformity. The hair of the horse is largely used by the Chaffinch, as is the fur of several other animals; but in the generality of nests the hairs of the cow predominate.

The texture of the nest is very strong, and, owing to the

nature of the materials, is very elastic, returning to its original shape even after severe pressure. Boys seldom take the eggs of the Chaffinch, because they are so plentiful; but they are too

NEST OF GOLDFINCH

apt to take the nest itself, knowing that it makes a safe and convenient basket for the eggs of rarer birds, and forgetting that they cause much sorrow to the poor birds that have spent so much trouble in preparing their home.

As I have already mentioned, there is some resemblance

between the nest of the chaffinch and that of the GOLDFINCH (*Fringilla carduelis*).

In point of beauty, neither yields to the other, for the materials are much the same, and the mode of structure is nearly identical. The nest of the Goldfinch, however, is shallower than that of the chaffinch, and the lichens and moss of which it is partly made are not stuck on the outside, but are woven so deeply into the walls that the whole surface is quite smooth.

The position of the two nests, however, is very different. Instead of choosing the forks of a bough, the Goldfinch likes to make its nest near the end of a horizontal branch, so that it waves about and dances up and down as the branch is swayed by the wind. It might be thought that the eggs would be shaken out by a tolerably sharp breeze, and such would indeed be the case, were they not kept in their place by the form of the nest. If one of the best examples be examined, it will be seen to have the edge thickened and slightly turned inwards, so that, when the nest is tilted on one side by the swaying of the bough, the eggs are still retained within. I have seen the branches of a tree violently agitated by ropes and sticks, and noticed that the eggs in a Goldfinch-nest retained their position until the branch was struck upwards close to the spot on which the nest was made, all the previous agitation having failed to dislodge them.

The lining of the Goldfinch's nest is unlike that which is used by the chaffinch. The latter bird mostly employs hair, while the former makes great use of vegetable-down, such as can be obtained from the willow, the coltsfoot, and other plants. Like other birds, the Goldfinch will not take needless trouble, and if it can find a stray tuft of cotton-wool, will carry it off, and work it into the nest. Sheep-wool is also used for the same purpose ; but the bird likes nothing so well as down, and will use it in preference to any other material. On this soft bed repose the five pretty eggs, white tinged with blue, and diversified with small greyish-purple spots. Now and then a small streak is seen; but the spots are the rule, and the streaks the exception.

Altogether, it is hardly possible to find a more beautiful group than is made by a pair of Goldfinches, their nest, and eggs.

THE nest of the BULLFINCH (*Pyrrhula vulgaris*) is unlike that of the goldfinch, though it is sometimes found in similar localities. This bird seems to be rather capricious in its ideas of nest-making, sometimes preferring trees, and sometimes building in shrubs.

There was a little spinney which I once knew, in which were any number of Bullfinch-nests, the underwood being very attractive to the birds. All the nests were built very low, seldom more than four feet from the ground, and, to the best of my recollection, were placed among the branches of hazel and dogwood. The nest of the Bullfinch is by no means so neat and smooth as that of the goldfinch, but is made in a much looser manner; the foundation being formed of slender twigs, usually those of the birch, and the inner wall of the nest woven of delicate fibrous roots. This wall is flimsy in structure, rather shallow, and neither so deep nor so round as that of the goldfinch. The lining is made of similar materials, but of a finer kind.

The quantity of sticks used as the foundation for this nest varies according to the kind of branch on which it is placed; for when the bird selects a forked twig, such as that of the hazel or dogwood, it uses a considerable quantity of sticks; but when it places its nest on the nearly horizontal spray of the fir, it finds a sufficient foundation ready made, and only just lays a few twigs to fill up a blank space. The egg of the Bullfinch is something like that of the goldfinch, but larger and more conspicuously spotted.

In some works upon the eggs and nests of birds, the Bullfinch is said to build in bushes of considerable height and size. Now, this is not necessarily the case, inasmuch as the spinney which has just been mentioned was composed entirely of trees and low brushwood, and the Bullfinches always preferred the latter. I certainly have often found their nests in tall bushes, and sometimes in trees ; but they were always placed at so low an elevation, that the height of the tree or bush had no effect on that of the nest.

THE bird next on our list is rather variable in its nesting.

The GOLDEN ORIOLE (*Oriolus galbula*) is seldom seen in England, and its nest even more seldom. Every year, however, a few stray nests are built in this country, as there are few years in which the journals devoted to natural history do not contain a notice of the bird being seen, and occasionally of its nest being found. In the warmer parts of the continent it is plentiful, and in Italy is regularly exposed in the markets towards the middle of autumn, when it has indulged in fruit for some time and has become very plump and fat.

In this condition it is well known to epicures under the name of Becquafiga, corrupted into Beccafico. It is not easily procured, as it is a very wary bird, and does not like to venture far from covert. In the autumn, however, its love of fruit conquers its fear of man, and it haunts the orchard in numbers, making no small havoc among the fruit. Even under such circumstances it is not easy of approach, and the gunner will seldom manage to secure his prey except by imitating its peculiar and flute-like notes. He must, however, be very careful in his mimicry, for the bird has a critical ear, and if it detects the imitator, is sure to slip through the foliage and fly off to its forest stronghold.

The nest of the Golden Oriole is always placed near the extremity of a branch, and in some cases is so constructed that it almost deserves to be ranked among the pensiles. It is always a pretty nest, and the accompanying illustration conveys a good idea of its general form. It is always more or less cup-like in shape, but the comparative depth of the cup is very variable, as in some cases it is scarcely deeper in proportion than that of the goldfinch, and rather saucer-shaped, while in others the depth even exceeds the width. Perhaps the nest may be altered in shape after the female begins to deposit her eggs, as is known to be the case with many birds, the additions being always made to the margin.

It is a remarkable fact that this enlargement of the nest should be common both to birds and insects. The reader may

GOLDEN ORIOLES AND NEST.

perhaps remember that the wasp, as well as other hymenoptera, lays an egg in the cell while it is yet shallow, and adds to the cell in proportion to the growth of the grub. The time of year, therefore, at which the nest of the Golden Oriole is found will have an influence on its shape, as the nest which is taken in the early spring, before the eggs are laid, will probably be shallower than that which is found in autumn, after the eggs have been hatched and the young reared.

The object for deepening the nest may probably be traced to the weather which happens to prevail. If the winds be light, the nest may remain in its flat and saucer-like form without endangering the safety of the eggs, but if the season should be inclement and tempestuous, a deeper nest is needed in order to prevent the eggs or young from being flung out of their home.

The body of the nest is formed chiefly of vegetable substances, usually the stems of different grasses, which are interwoven with wool, and thus made into a tolerably strong fabric. The female bird is said to be very affectionate, and to sit so closely on her nest that she will almost suffer the hand to be laid upon her before she will leave her post. In the illustration, the female bird is standing upright on the branch, and looking upwards, while the male is bending over the bough, and peering downwards, as if at some fancied foe. He can always be distinguished from his mate by the brighter gold of his plumage, the black spot between the eye and the beak, and the deeper black of his wings; whereas in the female, a tinge of blue invades the yellow, changing it to yellowish green, the wings are brown, edged with grey, and the black spot in front of the eye is altogether absent. Moreover, the breast and belly are marked with many longitudinal dashes of greyish brown.

ONE of the common American birds, the YELLOW-BREASTED CHAT (*Icteria viridis*) is not only remarkable for its really pretty nest, but for the manner in which it defends its home.

Although so chary of being seen that an experienced ornithologist may follow it for an hour by its voice, and never catch a glimpse of the bird, it is full of talk, and as soon as a human being approaches, it begins to vociferate reproaches in an odd series of syllabic sounds, which can be easily imitated. Mocking the bird is an unfailing method of doubling its anger, and will cause it to follow the imitator for a long distance, although it will under these circumstances keep itself hidden in the foliage. Wilson's account of the curious sounds which it utters is very graphic and interesting. 'On these occasions his responses are constant and rapid, strongly expressive of anger and anxiety, and while the bird itself remains unseen, the voice shifts from place to place amongst the bushes, as if it proceeded from a spirit. First is heard a repetition of short notes, resembling the whistling of the wings of a duck or teal, beginning loud and rapid, and falling lower and slower, till they end in detached notes. Then a succession of others, something like the barking of young puppies, is followed by a variety of hollow guttural sounds, each eight or ten times repeated, more like those proceeding from the throat of a quadruped than that of a bird; which are succeeded by others not unlike the mewing of a cat, but considerably hoarser.

'All these are uttered with great vehemence, in such different keys and with such peculiar modulation of voice as sometimes to seem at a considerable distance, and instantly as if just beside you; now on this side and now on that: so that, from these manœuvres of ventriloquism, you are utterly at a loss to ascertain from what particular spot or quarter they proceed. If the weather be mild and serene, with clear moonlight, he continues gabbling in the same strange dialect, with very little intermission, during the whole night, as if disputing with his own echoes.

'While the female is sitting, the cries of the male are still more loud and incessant. When once aware that you have seen him, he is less solicitous to conceal himself, and will sometimes mount up into the air, almost perpendicularly, with his legs hanging, descending, as he rose, by repeated jerks, as if highly irritated, or, as is vulgarly said, "dancing mad." All this noise and gesticulation we must attribute to his extreme affection for his mate and young; and when we consider the great distance from which in all probability he comes, the few young produced at a time, and that seldom more than once in the season, we can see the wisdom of Providence very manifestly in the ardency of his passions.'

The nest which the bird defends with such skill and courage is very well concealed in a dense thicket, and the bird is always best pleased if it can find a bramble-bush thick in foliage and well beset with thorns. Sometimes it is forced to content itself with a vine or a cedar, and in any case it is seldom more than four or five feet from the ground. The outer wall is made of leaves, within which is a layer formed of the thin bark of the grape-vine, and the line is formed of dried grasses and fibrous roots of plants.

Of our four British pigeons, two are branch-builders. The Stockdove places its nest in holes in trees, in holes in the ground, or on the tops of pollard oaks, willows, and similarly crippled trees. The Rockdove makes its rude nest in the crevices of the rocks which it frequents. But the Ringdove and the Turtle-dove are true branch-builders, and are therefore noticed in this place.

We will first take the Ringdove (*Columba palumbus*), sometimes called the Wood-pigeon, the Woodquest or queest, and the Cushat.

The nest of the Ringdove is placed in a variety of localities, for the bird is not in the least particular in this respect. Sometimes it is situated near the top of a lofty tree, and sometimes it is found in a hedge only a few feet from the ground. I have seen nests in both localities.

Mr. Waterton mentions a curious circumstance connected with this bird. In a spruce fir-tree there was the nest of a magpie, containing seven eggs, which were removed and those of the jackdaw substituted. Below this nest a Ringdove had chosen to fix her abode, and so the curious fact was seen, that on the same tree, in close proximity to each other, were magpies, jackdaws, and Ringdoves, and all living in perfect amity. It might have been supposed that the magpies and jackdaws would have robbed the nest of the Ringdove, but such was not the case. Moreover, the bird knew instinctively that she would not be endangered by her neighbours, for she came to the tree after the magpie had settled in it.

The nest of the Ringdove is of so simple a character as scarcely to deserve the name. The bird chooses a suitable spray, and lays upon it a number of sticks, which cross each other so as to make a nearly flat platform. Many birds make a similar platform as the foundation of their nest, but with the Ringdove it constitutes the entire nest. So slight is the texture

RINGDOVE AND NEST.

of the platform, that when the two white eggs are laid upon it they can be discerned from below by a practised eye, and it really seems wonderful that they can retain their position on such a structure.

Moreover, the open meshes of the nest allow the wind to blow freely between the sticks, so that nothing would seem to be more uncomfortable for the young. Above, they can certainly be sheltered by the warm body and protecting wings, but below they seem to be exposed to every blast. Yet they find shelter enough, and not only find it, but make it. With the generality of birds, the droppings are conveyed away by the parents, but with the Ringdove they are allowed to remain, when they rapidly fill up all the open interstices, and form a dry scentless plaster, which effectually defends the tender bodies of the young from the wind, and has the further effect of consolidating and strengthening the nest.

Although the nests are plentiful enough, and the eggs are common in the cabinet of oologists, it is not very easy to find a nest that is furnished with this curious plaster, probably because some one of the many foes which persecute the Ringdove has discovered the nest, stolen the eggs, or killed the parent before the young birds were hatched.

It has already been mentioned that, with many branch-building birds, the thickness of the nest, or of the platform on which it is placed, is regulated by the exposed or sheltered position of the branch, and such is the case with the Ringdove. Although in some instances, the platform is so flimsy that the eggs can be seen through the interstices, in other cases it is from half an inch to an entire inch in thickness. In all cases, the longest twigs are first laid, and followed by those of smaller size; and, although the whole structure is very rude, it is always made with sufficient care to assume a tolerably circular shape.

The Turtledove (*Columba turtur*) builds a nest of very similar form, and, if possible, even slighter in construction.

THE well-known WATER HEN OR MOOR HEN (*Gallinula Chloropus*) always places its nest near the water, but the bird seems to be very indifferent about the precise locality.

Sometimes it is made on the ground, and in that case is laid among sedges and rushes where the water cannot reach it. The Water Hen, however, is not averse to nesting in a warm and comfortable place, for Mr. Waterton mentions that on one occasion, when he had built a neat little brick house for a duck, and furnished it with dry hay for a nest, a Water Hen took possession of it, and the duck had to find a home elsewhere.

Sometimes the nest is made on a branch, and in that case the bird selects a very low bough which overhangs the water. I have found several nests thus placed, and in one case the only method of getting at the nest was to enter the water and

WATER HEN AND NEST.

swim round to it. It is a large and rudely made nest, and from its size appears to be more conspicuous than is really the case. When it is placed on a bough, the twigs of the same branch often dip into the water, and the nest looks like a bunch of weeds and other *débris* that have floated down the stream and been arrested by the branch.

The similitude is increased by a curious habit of the bird. When she leaves her nest, she pulls over her eggs a quantity of the same substances as those which form the materials of the nest, so that they are completely hidden from sight, and the form of the nest is quite obscured. It is true that the nest is not unfrequently found with the eggs exposed, but this apparent negligence is always caused by the frightened bird dashing off at the approach of the intruder, and having no time to cover her eggs properly. The object of covering the eggs was once thought to be the retention of heat, the neighbourhood of water being imagined to be injurious. As, however, many birds build as close to the water as does the Water Hen, and do not cover the eggs, it is evident that concealment and not warmth is the object to be attained.

I may mention that the illustration was sketched from a nest before it was removed, and that most of the nests have been drawn in the same manner from actual objects.

The eggs are many in number, seldom less than six, and often eight, and their united weight is far from inconsiderable, as they are fully proportioned to the size of the bird. The young are the oddest little beings imaginable, looking like spherical puffs of black down, rather than birds. They take to the water at once, and if the reader can manage to watch the mother and her little family, he will see one of the quaintest and prettiest groups that our country can supply. The little black balls swim about quite at their ease, keeping within a short distance of their parent, and traversing the water with a rapidity that reminds the observer of the *gyrini*, or whirligig beetles. In spite of the prolific nature of the bird, it is not so numerous as it might be, having many enemies in its youth, the worst of which is the pike, which comes up silently from below, opens its terrible jaws, and absorbs the unsuspecting bird.

CHAPTER XXIX.

FEATHERED BRANCH-BUILDERS—(concluded).

The SEDGE-WARBLER—Its nest and loquacity—The REED-WARBLER—Use of its peculiar tail—Localities haunted by the bird—Song of the Reed-Warbler—Its deep and beautifully balanced nest—Colour of the eggs—The INDIGO BIRD—The CAPOCIER—Familiarity of the bird—Le Vaillant's experiments—How the nest is made—Division of labour—Lovers' quarrels—Structure of the nest—Humming-birds again—The FIERY TOPAZ—Appearance of the nest—Its shape and the materials of which it is made—The HERMIT HUMMING-BIRDS and their nests—The RUBY-THROATED HUMMING BIRD—Variable dimensions of the nest—Concealment—Mr. Webber and his discoveries—Variable form and positions of the nest—Materials of which it is made—Its deceptive exterior—Feeding of the young—The RED-BACKED SHRIKE—Use of the Shrike in falconry—Their singular mode of feeding—Impaled prey—Conspicuous character of the nest—Popular ideas concerning the Red-backed Shrike—Structure of the nest.

ANOTHER bird that loves to build near water is the pretty little SEDGE WARBLER (*Salicaria phragmitis*).

The nest of this bird is placed at a very low elevation, usually within a foot or so from the ground, and raised upon rushes, reeds, or other coarse herbage, which is found abundantly in such places. There is more material in the nest than might be supposed from the size of the bird and the slender stems by which it is supported. Viewed from the exterior, it seems to have the ordinary cup-shaped form which is so prevalent among small birds, but looked at from above, the apparent depth is seen to be owing to the mass of material, the hollow being singularly small and shallow. It is a well-made nest, the general framework being formed of leaves of grass-blades, while strength, warmth, and density are attained by the quantity of wool and hair which are woven into the fabric.

The Sedge Warbler is well known for its loquacity, and its ceaseless chatter. Should it be silent, a stone flung among the reeds and sedges will always induce it to recommence its little song.

A REMARKABLY beautiful nest is built by one of the British birds, but is not often found, on account of the localities where it is placed.

The architect of this nest is the REED WARBLER (*Salicaria* [or *Curruca*] *arundinacea*). It is a pretty little bird, bright brown above, yellow-brown below. In some respects it resembles the sedge warbler, but does not possess the remarkable wedge-shaped tail of that bird. R. Mudie, in his History of British Birds, offers the following suggestion respecting this difference of form. When treating of the sedge warbler, he remarks that the slender head, pointed bill, and wedged-shaped tail are useful to the bird by enabling it to glide between the tall aquatic plants among which it resides and finds its food. Of the Reed Warbler he writes as follows :—

'That the bird is not adapted for so many situations as the sedge bird, might be inferred from the different form of the tail, which is more produced and not wedge-shaped, so that while it answers better as a balance on the bending reeds or other flexible aquatic plants, it would not be so convenient among the unyielding sprays of a hedge or brake. The bird rarely, if ever, perches upon the tops of reeds, even on its first arrival, and when the song of invitation to a mate is given, its place is on a leaf or a leaning stem, though upon an emergency it can cling to an upright one, the stiff feathers of the tail acting as a sort of prop.

'It is not easily raised, and remains but a very short time upon the wing, but it is by no means timid on its perch, upon which, if it be very flexible, it sits with its wings not quite closed, but *recovered*, so as to have a little hold on the air, and thereby either prevent its fall or be ready when a gust comes, to bear it to a more secure footing. Its food is found wholly over the stagnant waters. The Reed Warbler does not come until the reeds are considerably advanced, and it departs before they are cut; so that it dwells in peace, and especially in the mornings about the end of May and the beginning of June it may be observed with the greatest ease.'

Still, although the bird be common, and although it is bold enough to admit of approach, it is not generally familiar, simply because none but professed naturalists are likely to look for it in the spots which it frequents. The Reed Warbler loves a large patch of marshy land almost wholly covered with stagnant water, and full of the reeds among which its home is made. Such a place is not agreeable to the pedestrian, for although an hour spent in wading through water knee-deep is no difficult or even unpleasant task, yet no one likes to meet also with mud of various and unknown depths, as is the case in the great reed swamps where the birds most love to build. Even the song of the Reed Warbler does not attract attention. Though musical in tone, it is very feeble in power and monotonous in character, consisting of several hurried notes in a low warble, which can only be heard at a little distance.

The nest of this bird is supported between three or four reeds, and is remarkably deep in proportion to its width. The object of this depth is evident. To bend as a reed before the wind is a proverbial saying, and any one who has seen a large mass of reeds on a stormy day must have been impressed with their graceful curves. As the blasts of the wind pass over them, they bend in successive waves like the billows of the sea, and are sometimes bowed so low that their tips nearly reach the water.

A nest, therefore, which rests on such pliant supports must be thrown out of its perpendicular by every breath of wind, and unless it were very deep the eggs would be flung out. The great depth, however, of the nest counteracts the deflection of the reeds; and, however fiercely the storm may rage, the Reed Warbler sits securely in her nest, even though it be sometimes nearly bowed to the surface of the water. The materials of the nest are generally taken from the immediate neighbourhood, the body of the nest being composed of broken rushes and moss bound together with reed leaves, and the lining made almost wholly of cows' hair.

MANY foreign birds are excellent branch-builders.

IN Southern Africa there is a small, simply coloured, but interesting bird, called by Le Vaillant the CAPOCIER (*Drymoica maculosa*) because it builds in a cotton-yielding tree, called by the Dutch colonist Capoc-bosche.

The attention of that naturalist was directed to the bird in the following manner.

Being, in common with all true naturalists, a lover of birds in their living state, and being in no wise disposed to kill them without necessity, he had contrived to tame a pair of little brown birds, which at last became so familiar that they would enter his tent. On these terms they remained until the beginning of the breeding season, when they began to come less regularly, and then to absent themselves for several successive days. About this time they became thieves. M. Le Vaillant was accustomed to keep on his table a quantity of tow and cotton-wool, which he used in stuffing and otherwise preparing the skins which he had procured for his collection. The birds seemed suddenly to take a wonderful fancy to the tow and cotton-wool, and were continually flying off with them, sometimes stealing a piece that was nearly as large as both the birds together.

Struck with this sudden fancy of the birds, Le Vaillant determined to watch them, and soon traced them to a capoc-bosche tree which grew at some distance. and in a remarkably retired spot. Among the branches of this tree they had already begun their nest, which consisted of a quantity of moss pressed tightly into the forks of a bough, and which was at the time only in a rudimentary condition. The moss, in fact, was the foundation of the nest, upon which the beautiful walls were intended to be built, just as in the habitation of many other birds there is a foundation of substances more solid than the materials of which the walls are made.

Into this nest the Capociers were weaving the stolen stores of cotton-wool, working it in a manner that will be presently described. Le Vaillant soon discovered that the legitimate substance of the nest-walls was the soft, white down produced by certain plants, and that the birds used an enormous amount of materials in comparison with their own size. As, however, they found that upon the naturalist's table was always a plentiful supply of vegetable down and fibres ready plucked, they ingeniously saved themselves the trouble of collecting, and simply resorted to the hospitable tent.

The male was the principal collector of materials, and the female the chief architect. He used to fly off, and return with a mass of cotton-wool, moss, or tow, and deposit it close to the spot where his mate was at work. Then she would take the materials, arrange them, press them into form, and only ask his assistance in carrying out her plans. He pressed, and pecked and pulled the cotton-wool so as to reduce it to a kind of felt, but did not seem to originate any architectural ideas, leaving them to his more ingenious mate.

Le Vaillant's account of the mode of working is so interesting and elegant that in justice to himself it must be given in his own words. After describing the process of fetching materials and laying them in their places, he proceeds as follows :—

'This agreeable occupation was often interrupted by innocent and playful gambols, though the female appeared to be so actively and anxiously employed about her building as to have less relish for trifling than the male, and she even punished him for his frolics by pecking him well with her beak. He, on the other hand, fought in his turn, pecked, pulled down the work which they had done, prevented the female from continuing her labours, and, in a word, seemed to tell her, "On account of this work you refuse to be my playmate, therefore you shall not do it."

'It will scarcely be credited that, entirely from what I saw and knew respecting these little altercations, I was both surprised and angry at the female. In order, however, to save the fabric from spoliation, she left off working, and fled from bush to bush, for the express purpose of teasing him. Soon afterwards, having made matters up again, the female returned to her labour, and the male sang for several minutes in the most animated strains. After his song was concluded, he began again to occupy himself with the work, and with fresh ardour carried such materials as his companion required, till the spirit of frolic again became buoyant, and a scene similar to that which I have described occurred. I have witnessed eight interruptions of this kind in one morning. How happy birds are! They are certainly the privileged creatures of nature, thus to work and sport alternately, as fancy prompts them.

'On the third day the birds began to rear the side walls of the nest, after having rendered the bottom compact by repeatedly pressing the materials with their breasts, and turning themselves round upon them in all directions. They first formed a plain border, which they afterwards trimmed, and upon this they piled up tufts of cotton, which was fitted into the structure by beating and pressing it with their breasts and the shoulders of their wings, taking care to arrange any projecting corner with their beaks, so as to interlace it into the tissue, and to render it more firm. As the work proceeded, the contiguous branches of the bush were enveloped in the side walls, but without damaging the circular cavity of the interior. This part of the nest required many materials, so that I was quite astonished at the quantity which they used.

'On the seventh day their task was finished, and, being anxious to examine the interior, I determined to introduce my finger, when I felt an egg that had been probably laid that morning, for on the previous evening I could see that there was no egg in it, as it was not then quite covered in.

'This beautiful edifice, which was as white as snow, was nine inches in height on the outside, whilst in the inside it was not more than five. Its external form was very irregular, on account of the branches which it had been found necessary to enclose; but the inside exactly resembled a pullet's egg placed with the smaller end upwards. Its greatest diameter was five inches, and the smallest four. The entrance was two-thirds or more of the whole height as seen on the outside, but within it almost reached the arch of the ceiling above.'

One of the most remarkable points of this singularly beautiful nest is the firm texture of the walls. Externally, the nest looks as if it were a mere large hollow bunch of cotton-wool with a hole near the top, and seems to be so fragile that the eggs would fall through the fabric. But when the inside of the nest is viewed, it is seen to be composed of a kind of felt, as firm and close as if it had been formed by human art, so that neither wind nor wet can penetrate; and it is capable of upholding a much greater weight than would be introduced into it. To pull out a tuft of the cotton-wool is impossible without tearing a hole in the fabric, so closely are the delicate fibres interwoven with each other.

In the accompanying illustration are shown the nests of two species of Humming Bird.

The oddly-shaped nest which occupies the upper part of the drawing is made by the Fiery Topaz (*Topaza pyra*), one of the most magnificent of these lovely birds. Indeed, Prince Lucien Buonaparte calls it the most beautiful of the Trochilidæ, and it is hardly possible to imagine a bird that can surpass it in brilliancy. The body is fiery scarlet, the head velvet-black, the throat glittering emerald, with a patch of crimson in the centre; the lower part of the back is also green, and the long, slender, crossed feathers of the tail are purple with a green gloss. So magnificent a bird can have but few rivals, and there is only one species which even approaches it in beauty. This is the Crimson Topaz (*Topaza pella*), a bird which is nearly allied to it, and which much resembles it in general colouring. It may, however, be distinguished by the colour of the body, which is crimson instead of scarlet.

The nest which is built by the Fiery Topaz is really a wonderful structure.

Its shape is remarkable, and is well shown in the illustration. It is fastened to the branch with extreme care, as is clearly necessary from its general form. The most curious point about the nest is, however, the material of which it is made. When it was first discovered no one knew how the bird could have built so strange a structure. It looked as if it were made of very coarse buff leather, and was so similar in hue to the branches that surrounded it, that it seemed more like a natural excrescence than a bird's nest. The reason for this similitude was simple enough. It was made of a natural excrescence, and therefore resembled one.

When the Fiery Topaz wishes to build a nest, it goes off to the trees, and searches for a kind of fungus belonging to the

FIERY TOPAZ AND HERMIT.

genus boletus, and with this singular material it makes its home. It is tough, leathery, thick and soft, and in some curious manner the bird contrives to mould the apparently intractable substance into the shape which is represented in the illustration. The non-botanical reader may form an idea of the appearance of the nest, by supposing it to be made of German tinder, which is, in fact, a kind of boletus which has been pressed, dried, and steeped in a weak solution of nitre.

The lower figure in the same illustration represents the nest of another Humming Bird (*Phaethornis eurynome*), belonging to the pretty little group which are popularly called Hermits, and which may be recognised by the peculiar shape of the tail, which is regularly graduated, the two central feathers being, however, much longer than the others. They are inhabitants of Venezuela.

All the Hermits are remarkable for the beauty of their homes, and the present species is mentioned as affording a good example of nest-making. The nest is always long and funnel-shaped, and is hung either to a leaf or the delicate twig of a tree, according to circumstances. The materials of which the nest is made are rather various, consisting of vegetable fibres, especially those downy, cotton-like filaments which are furnished by so many plants, of small herbs, and spider webs. The last-mentioned substance is employed for the purpose of binding the materials together, and is used also in fastening the nest to the support on which it hangs.

THERE is another species of this beautiful group, called the RUBY-THROATED HUMMING BIRD (*Trochilus colubris*), which is generally accepted as the typical species. This lovely bird is plentiful in many parts of America, and is sometimes seen as far North as Canada. It derives its popular name from the feathers of the throat, which glitter as if made of burnished metal, and glow with alternate tints of ruby and orange. The general colour of the body is green, and the wings are purple-brown. The two sexes are coloured after the same manner, with the exception of the ruby gorget, which only belongs to the male, and which is not attained until the second year. There is no species more common in museums and ornamental cases than this, because it is as plentiful as it is lovely. That it should be plentiful, or indeed that any species of Humming Bird should be anything but scarce, is matter of wonder, inasmuch as they never lay more than two eggs, and in all probability do not rear more than three, or perhaps four, young in the course of a season.

The general habits of this tiny bird are well worthy of notice, but at present we must content ourselves with it as it appears in its nest-making capacity. Being a very small bird, only three inches and a half in total length, and very slenderly made, the nest is necessarily small. But, although we so often find that little birds build large nests, we cannot but notice that the nest of this Humming Bird is even smaller than the size of its occupant seems to require. It is round, neatly made, and has thick walls and a small hollow.

The bird has a wonderful power of concealing the nest, which cannot be discovered except by a practised nest-hunter, so closely does it resemble a knob upon a branch. So careful, too, is the female of her home, that she does not fly straight to it, but rises high in the air, and then darts down among the branches with such rapidity that the eye cannot follow her movements, and she is fairly seated in her nest before the spectator knows exactly in which direction she has gone.

This curious trait seems to have been discovered by Mr. C. W. Webber. He had successfully tamed some Ruby-throats, and determined to find a nest, so that he might obtain the young. After finding that a pair of Humming Birds had been seen near a certain spot on a river, he set himself determinately to discover the nest. By degrees they were watched to a point of the river, but there they always disappeared, as they had a habit of shooting perpendicularly into the air until their tiny bodies were lost to sight. At last, however, the patient watchfulness of the observer was rewarded by catching a glimpse of the female bird, as she descended perpendicularly from the height to which she had risen, and in this manner was the nest discovered.

The same agreeable writer relates an anecdote respecting the discovery of a nest belonging to the Emerald-throated Humming Bird, an edifice which is very similar to that which is made by the Ruby-throat. He had been in vain looking for a nest, when 'chance favoured me somewhat strangely about this time. I had been out squirrel-shooting early one sweltering hot morning, and on my return had thrown myself beneath the shade of a thick hickory, near the bank of a creek. I lay on my back, looking listlessly out over the stream, when the chirp of the Humming Bird and its darting form reached my senses at the same instant. I was sure I saw it light upon the limb of a small iron-wood tree, that happened to be exactly in the line of my vision at that instant.

'In about five minutes another chirp and another bird darting in. I saw this one drop upon what seemed to be a knob or an angle of the limb. I heard the soft chirping of greeting and love. I could scarcely contain myself for joy. I would have given anything in the world to have dared to scream, "I've got you, I've got you at last!" By a great struggle I choked down my ecstasy and kept still. One of them now flew away, and after waiting fifteen minutes, that seemed a week, I rose, and with my eyes steadily fixed on that important limb, I walked slowly down the bank, without, of course, seeing where I placed my feet.

'But the highest hopes are sometimes doomed to a fall, and a fall mine took with a vengeance! I caught my foot in a root, and tumbled head foremost down the bank into the river! I suppose that such a ducking would have cooled the enthusiasm of most bird-nesters, but it only exasperated mine. I shook off the water, and vowed that I would find that nest if it took me a week. But how to begin was the question, for I had lost the limb, and how was I to find it among a hundred others just like it?

'The knot that I had seen was so exactly like other knots upon other limbs all round it, that the prospect of finding it seemed a hopeless one; but, "I'll try, sir," is my favourite motto. I laid myself down as nearly as possible in the position which I had originally occupied, but, after some twenty minutes' experiment, came to the conclusion that my head had been too much confused by the shock of my fall and ducking for me to hope to make much out of this method. Then I went under the tree, and commencing at the trunk, with the lowest limb which leaned over the water, I followed it slowly and carefully with my eye out to the extremest twig, noting carefully everything that seemed like a knot. This produced no satisfactory result after half an hour's trial, and with an aching neck I gave it up in despair, for I saw half a dozen knots, either one of which seemed as likely to be the right one as the other.

'I now changed my tactics again, and, ascending the tree, I stopped with my feet upon each one of those limbs and looked down along it. It was a very tedious proceeding, but I persevered. Knot after knot deceived me, but, at last, when just above the middle of the tree, I caught a sharp gleam of gold and purple among the leaves, and, looking down upon the last limb to which I had climbed, almost lost my footing for joy, as I saw, about three feet out from where I stood, the glistening back and wings of the little bird just covering the top of one of these mysterious knots that was about half the size of a hen's egg.

'The glancing head, long bill, and keen eyes were turned upwards, and perfectly still, except the latter, which surveyed me from head to foot with the most dauntless expression. It seemed not to have the slightest intention of moving, and I would not have disturbed it for the world. It was sufficient to me to gaze on my long-lost treasure. Its pure white breast—or throat rather, for the breast was sunk in the nest—formed such a sweet and innocent contrast with the splendour of its back, head, and wings.' The capture of the little birds which were afterwards hatched in that nest served to set at rest the question of the Humming Bird's food. They lived mostly on syrup, but were obliged to fly off and eat the tiny garden spiders as they lay in the middle of their radiating webs.

The nest of the Ruby-throated Humming Bird seems to be rather variable in form and material and situation, but has always a peculiar character which enables the experienced observer to recognise it. According to Wilson, it is sometimes fixed on the upper part of a horizontal branch, as was the case with the nest so graphically described by Mr. Webber. Sometimes it is seen actually upon the trunk of a tree, attached to the bark by its side; and in a few rare instances it has been found in a garden, attached to some strong-stalked herb. Generally, however, the bird selects a white oak sapling if it builds in the woods, and a pear-tree if it prefers the garden.

The tiny nest is scarcely more than one inch in width and the same in depth, so that its size is very small when compared with that of its occupants, which, when full grown, are more than three inches in total length. The materials of which the nest is made are principally the delicate cotton-like fibres which form the 'wings' of certain seeds, such as those of the thistle, and are so carefully woven together that they form a tolerably stout wall. Upon this wall are stuck quantities of a light grey lichen which is found on old fences and trees, so that the external appearance of the nest is rendered very similar to that of the branch on which it is placed. The lining is composed of the fine hairs which clothe the stalks of mullein and ferns and other pubescent plants, and forms a thick, soft bed on which repose the two minute pearly eggs.

The nest is not merely placed upon the branch, because in that case it would present a decided outline, and be comparatively easy of recognition. On the contrary, the base of the nest is partly continued round the branch, so that the whole fabric rises gradually from the bough, as if it were a natural excrescence.

When the young are hatched they are fed by thrusting their beaks into the opened mouths of their parents, and extracting the supply of liquid sweets which have been collected from the flowers.

In the hedgerows of our own country may often be found a nest which is not only pretty in itself, but remarkable for its accessories. This is the home of the Red-backed Shrike (*Enneoctonus collurio*).

The predatory habits of the Shrikes are well known, and one species, the Great Grey Shrike (*Lanius excubitor*), was formerly used as a falcon for the purpose of catching winged game. True, the bird was not considered as a veritable hawk, and in the old days of sumptuary laws, when each degree of rank had its own particular species of hawk, this was a fact of some significance, showing that those who thus employed the Shrike were not of gentle blood.

The popular notion of the time supplied another reason why the Shrike was looked upon with disdain as a bird-catcher. It was supposed to use guile in securing its prey, instead of openly conquering in fair chase. 'Sometimes,' writes an old sporting author, 'upon certain birds she doth use to prey, whome she doth entrappe and deceive by flight, for this is her desire. She will stand at pearch upon some tree or poste, and there make an exceeding lamentable crye and exclamation, such as birds are wonte to do, being wronged or in hazard of mischiefe, and all to make other fowles believe and thinke that she is very much distressed and stands in need of ayde; whereupon the credulous sellie birds do flocke together presently at her call and voice, at what time if any happen to approach neare her she out of hand ceazeth on them, and devoureth them (ungrateful subtill fowle!) in requital for their simplicity and pains.

'Heere I end of this hawke, because I neither accompte her worthy the name of a hawke, in whom there resteth no valour or hardiness, nor yet deserving to have any more written upon her propertie and nature. For truly it is not the property of any other hawke, by such devise and cowardly will to come by their prey, but they love to winne it by main force of wings at random, as the round winged hawkes doe, or by free stooping, as the hawkes of the tower doe most commonly use, as the falcon, gerfalcon, sacre, merlyn, and such like.'

The Shrikes have a peculiarity which is not shared by any other predacious bird. When they have slain their prey, no matter whether it be bird, beast, reptile, or insect, they take it to some thorn tree, and there impale it, pressing a long and sharp thorn into the body, so as to hold it firmly. The Great Grey Shrike will thus transfix the smaller birds, frogs, field-mice, and other creatures which are nearly as large as itself, and in some instances it has been known to kill and impale the thrush. It does not always employ thorns for this purpose, but will use sharply-pointed splinters of wood, or even an iron spike if no better instrument can be found.

Why it should have recourse to such a singular mode of holding its prey is quite a mystery. Some have said that the digestive organs of the Shrike are incapable of dissolving fresh meat, and that the bird is obliged to render its prey semi-putrid by exposure before it can venture to make a meal. But, as the Shrike frequently eats a little bird or insect as soon as it is caught, this theory falls to the ground.

Whatever theory may be right or wrong, the fact remains that the Shrikes impale the creatures which they have killed, and prefer to hang them near their nests. The Red-back Shrike makes insects its chief prey. The nest of this Shrike always affords a curious sight, and as the bird is plentiful it may easily be seen.

There is not the least difficulty in finding a Shrike's nest, for the owner really seems to use every means which can attract attention. In the first place, it is a bird of insatiable curiosity. It will follow, or rather precede, a human being for half an hour at a time, keeping always some thirty or forty yards in front, settling near the top of a hedge, and wagging its long tail up and down as if to make itself more conspicuous. Last year I amused myself by making a Shrike move up and down a long hedge for a very long time, while I was insect-hunting among the flowers. Whenever the Shrike begins to act in this manner, it may generally be presumed that a nest is at no great distance.

Then, if perchance the careful observer should note these signs and approach the spot where the nest is placed, the bird sets up a hideous squall, just as if it intended to inform the searcher that he was right at last. The alarm cry of the blackbird is quite enough to draw attention as the bird flies through the underwood; but at all events it is only a short cry, and the bird is soon out of sight; but the Shrike remains on or near the nest while it continues to utter its harsh screams, and flies away noisily when the intruder is close at hand.

The nest itself is large, and not concealed with any care,

while around it are stuck humble bees, cocktail beetles, ground beetles, and a variety of other insects, each impaled upon a thorn, and forming admirable indications to the nest-hunter. Sometimes, but seldom, young birds are impaled instead of insects, and in such cases they are always callow nestlings, and are fixed by a thorn run between the skin and the flesh, instead of being pierced through the body, as is the method employed with insects.

There is a popular idea that the bird always has *nine* impaled creatures at hand, and that when it eats one it catches another, and with it replaces the one which has been eaten. In consequence of this notion, which prevails through several countries, the bird is called Nine-killer. The generic name, Enneoctonus, is composed of two Greek words which have a similar signification. So strongly is this idea held by some persons, that I have seen a treatise upon instinct, where the Shrike was gravely produced as an example of arithmetical powers possessed by birds. These theories generally fail when confronted by facts. I have seen numberless Shrikes' nests; and, though in some cases there may have been nine impaled animals, in some there were more and in others less.

The nest itself is neatly, though loosely, built of roots, moss, wool, and vegetable fibres, and is lined with hair. I have mostly noticed it about five feet from the ground; and, although it is said to be closely hidden, have always found it a peculiarly conspicuous nest.

CHAPTER XXX.

BRANCH-BUILDERS.

SPIDERS AND INSECTS.

Remarkable Spider Nests in the British Museum—Seed-nests and Leaf-nests—Nest of the ICARIA—The equal pressure and excavation theories—Nest of MISCHOCYTTARUS and its remarkable form—Nest of the RAPHIGASTER—Summary of the Argument—The PROCESSIONARY MOTH—Reasons for its name—How the larvæ march—Damage done by them to trees—The social principle among Caterpillars—Mr. Rennie's experiments—The LACKEY MOTH—Supposed derivations of its popular name—The eggs, larvæ, and perfect insects—Habits of the Moth—The BROWN-TAILED MOTH—Locality where it is found—Its ravages abroad—The APOICA and its remarkable nests—Moth Nests from Monte Video.

WE have already seen several nests built by SPIDERS, some of which are made in the earth, others are strictly pensile, and others may fairly come into the present group. The specimens from which the drawings were made are in the collection of the British Museum, some in the upper and others in the lower rooms. Of the architects, the manner in which the nests were made, and the reasons why they were so singularly constructed, I can say nothing, because no record is attached to the specimens. Still, they are so curious that they have found a place in this work, and it is to be hoped that the very fact of their publicity will induce travellers to search for more specimens and to describe their history.

Differing as they do in shape, colour, and material, they have one object in common, namely, the rearing of the young. They are clearly nests in the true sense of the word, being devoted not to the parents, but to the offspring. At the upper part of the illustration may be seen a number of long, spindle-shaped bodies, suspended from a branch. These are drawn about half the full size, in order to allow other specimens to be introduced into the same illustration for the purpose of comparison. In colour they are nearly white, with a slight yellowish tinge, and are very soft and delicate of texture, so that when viewed in a good light they form a very striking group of objects.

Immediately below these nests may be seen a singular-looking object. which few would recognise as the work of a spider. Such,

PENSILE SPIDER'S NEST.

however, is the case, the creature being urged by instinct to take several concave seed-pods, and to fix them together, as seen in the drawing. The seed-pods are fastened firmly together with the silken thread of which webs are made, and in the interior the eggs are placed. The drawing is reduced about one-third in proportion to the actual object. Several of these singular nests are in the collection at the British Museum.

Occupying the lower part of the illustration is seen a leaf upon which are piled a number of fragments of leaves, so as to form a rudely conical heap. This is also the work of a spider, and is made with even more ingenuity than the two preceding specimens. In the first instance, the spider has spun a hollow case of silk, similar in principle of construction, though not in form, to the spherical egg cases made by several British spiders. In the second instance, the creature has chosen a number of concave seed-pods, and, by adjusting their edges together and fastening them with silk, made a hollow nest, which only requires to be lined in order to make it a fit nursery for the young. But, in the present example, the work of nest-making has been much more elaborate, for the structure has been regularly built up of a great number of pieces, each being arranged methodically upon the other, very much as children in the streets build their oyster-shell grottoes. The labour must have been considerable, even if the spider had nothing to do but to arrange and fasten together pieces of leaves which had already been selected.

In the accompanying illustration three most remarkable nests are given, all of them the work of hymenopterous insects, and all serving in some degree to illustrate the hexagonal system of cell-building, so common among the hymenoptera.

Of these, perhaps, the central figure is the most interesting, because it entirely sets at rest a question which is periodically agitated. It is made by an insect belonging to the genus *Icaria*. Perhaps my readers may remember that on a former page the celebrated bee-cell problem is described, and that mention is made of the many theories which have been invented to solve the riddle. Among them the two most conspicuous are those which are known as the equal pressure theory and the excavation theory. Differing as they do in many respects—one attempting to prove that each cell is forced into the hexagonal shape by the pressure of six cells surrounding it, and the other that the cell is made hexagonal by the cutting away of material from six surrounding cells—they both agree in one point, namely, that the normal shape of the cell is cylindrical, and that it only assumes the hexagonal form by mechanical means.

MISCHOCYTTARUS. ICARIA. RAPHIGASTER.

These questions were briefly mentioned, because an entire omission of them would appear negligent, but they were not followed up because the nests that would set them at rest belonged to another group. We will first take the central nest.

The specimen from which this was drawn was fortunately in an unfinished state, only eight cells being made, and some of these but partly finished. As the reader may see by reference to the illustration, all the cells are hexagonal, whether finished or incomplete, and moreover, that the edges of the hexagon are quite sharp and well defined.

Now, if either of the two theories were true, the cells would not have assumed this shape. Where are the six surrounding cells that are needed to compress the outermost cell into an hexagonal? Or where are the six surrounding cells from which the hexagon was excavated? There are none. The outermost cell, for example, is perfectly free on five of its sides, being only attached to the neighbouring cell by the sixth side. Compression, therefore, has not been employed, because there is nothing that can compress it; neither has excavation been used, because there is no material to be excavated. No one, on looking at this group of cells, can deny that the hexagonal form is produced by the direct labours of the insect, and not by any secondary mechanical means.

Perhaps some one who has not examined the actual object might say that the materials of which the cells are made are sufficiently stiff to need no support of contiguous cells. Now the substance of this remarkable nest is singularly slight, the walls being not thicker than the paper on which this account is printed, and the material is quite soft, as may be seen by the curvature produced by the mere weight of the structure. Yet none of the cells are united by more than three sides, the greater number by two only, and the external cells merely by a single side, leaving five sides and four angles perfectly free.

In this particular specimen the material has evidently been varied, the insect having been forced to employ different substances in forming its home, as is seen by the pale and dark rings alternately surrounding the cells. The insect which makes this curious home is of moderate size, and is greyish-black, banded with yellowish-white. The abdomen is tolerably stout and sharp-pointed, and is attached to the thorax by a short brownish footstalk. This insect is a native of Natal. Other species of the same group will be mentioned in the course of the following pages.

In the left-hand upper corner may be seen a very remarkable triple nest depending from a branch. This is the work of an insect called *Mischocyttarus labiatus*, which belongs to the family Polistidæ. Like the nest of the preceding insect, it is attached to the bough by a slender and tolerably long footstalk, and the mouths of the cells are downwards, as is always the case with these insects.

Generally, the group of cells is single, but occasionally a more perfect nest is found, which, like the specimen figured in the illustration, has three distinct cell groups, each pendent from the centre of the group above. This may seem rather a dangerous method of suspending the nest, but it is not more so than that which is employed by the common wasp, which builds tier under tier of cells, hanging each tier from its immediate predecessor by little pillars of the same paper-like material as that of which the cells are constructed; or very much, indeed, as the roadway of a suspension bridge is hung from its arch instead of being placed upon it. The insect itself is smaller than the preceding, and is almost uniformly brown.

The last of these three groups is particularly entitled to notice, on account of its bearing upon the hexagonal principle, which has been so often mentioned. The name of the insect is Raphigaster Guiniensis, and, as its name implies, it is a native of Western Africa.

The nest consists of a group of long cells, and suspended from a footstalk. The material of which the nest is composed is peculiarly soft and flimsy, reminding the observer of the worst and most porous French paper. The cells are so thin that the light shines through their delicate walls, and they are so soft that they yield to the least pressure. Each cell is small at the base, and increases regularly towards the mouth, like a reversed sugar-loaf.

Now, if the real cause of the hexagonal form were to be found in the equal pressure of surrounding cells, the central cells of this group ought to be hexagons, for they are soft, pliable, and their conical form renders them peculiarly liable to be squeezed out of shape. Yet, on examining the nest, we find that all the cells retain their conical form, the central cells being as rounded as those on the exterior, and their mouths being as circular.

These examples entirely destroy both theories.

In the first instance we have nests of which the cells are perfectly hexagonal throughout, although some of them are only attached by one side, and are not pressed upon at either of the five remaining sides. We find that the external angles are as sharp, and their internal measurement as true, as those which occupy the very centre of the bee-comb ; so that pressure is clearly not the cause of the hexagon. That excavation is not the cause is also evident, from the fact that the external cells cannot have been excavated, and yet are hexagonal.

These examples, therefore, show that the hexagonal form can exist without pressure. But, as if to show that pressure can exist without producing the hexagonal form, we have the nest of the Mischocyttarus, whose long, delicate, soft-walled cells are grouped round each other, and yet retain their conical form, so that at any part of them a transverse section would show a circular edge.

The insect which makes this nest is rather long, measuring perhaps an inch in length. The colour is pale yellow, and the abdomen is much elongated, and attached to a slender footstalk or peduncle nearly as long as itself. Several of the cells have been occupied by larvæ which have begun to assume the pupal condition, as is shown by the white covers over their mouths.

ONE of the most remarkable of these branch-building insects is that which has been appropriately named the PROCESSIONARY MOTH (*Cnethocampa processionea*). This curious moth lays a number of eggs, mostly upon the oak, and as soon as they are hatched the little creatures begin to form their home.

Externally it is not unlike that of the brown-tailed moth, but it differs in one respect, namely, that it is not divided into separate chambers, and has only one aperture. When the larvæ sally out for the purpose of procuring food, they spin guide lines, as is the case with many other caterpillars. But, instead of going out singly into the world, each to find its own food in its own way, they march out in regular order, like a military party on a foraging expedition.

A single caterpillar is always the leader, and often is followed by one or two others in Indian file. Presently, however, the caterpillars march two deep, and, if a large number should be on the move, the line is sometimes from five to six deep. They are all very close to each other, so that the procession flows on in one unbroken line, and until the observer is close to it, he cannot see that its component parts are moving at all.

The reader may remember that two species of wasp, namely Vespa vulgaris and Vespa germanica, will work harmoniously at the same nest. This curious sociability, which is contrary to the usual custom of nature, is shared by moths as well as wasps. When experimenting upon the nests of this species, M. Réaumur found that several distinct broods of caterpillars would spin a common web and live in peace together, just as if they had been the offspring of one mother. Mr. Rennie, however, carried the experiments still farther, and found that two different species would act in the same social manner.

'We ourselves ascertained during the present summer (1829) that this principle of sociality is not confined to the same species, or even to the same genus. The experiment which we tried was, to confine two broods of different species to the same branch, by placing it in a glass of water to prevent their escape. The caterpillars which we experimented upon were several broods of the brown-tail moth (*Porthesia auriflua*) and the lackey (*Clisiocampa neustria*). These we found to work with as much industry and harmony in constructing their common tent as if they had been at liberty in their native trees; and when the lackeys encountered the brown tails they manifested no alarm nor uneasiness, but passed over the backs of one another as if they had made only a portion of the branch.

'In none of their operations did they seem to be subject to any discipline, each individual appearing to work in perfecting the structure from individual instinct, in the same manner as was remarked by M. Huber in the case of the hive bees. In making such experiments, it is obvious that the species of caterpillars experimented with must feed upon the same sort of plant.'

One remark ought to be made on this interesting narrative. The author lays some stress on the fact that the two insects belonged not only to different species, but to different genera. It must, however, be remembered that although the distinction of insects into species is easy enough, their grouping into genera is quite arbitrary, depending entirely on the classifier. Linnæus, for example, divided all the butterflies into two genera, while the modern classification admits some thirty genera. While, therefore, we may lay every stress on the species, we need not trouble ourselves much about the genus.

The two moths mentioned in this history are very different in appearance, and the larvæ are still more unlike. They have, however, this point of similarity, that they construct large dwellings upon branches, spinning them of silk, and making them large enough to contain a whole brood at once. The Lackey moths are so called on account of the bright colours of the caterpillars, which are striped and decorated like modern footmen. Some species, however, derive the name from a different source.

When the mother insect lays her eggs, she deposits them on a small branch or twig, disposing them in a ring that completely encircles the twig, as a bracelet surrounds a lady's wrist. When she has completed the circle, she covers the eggs with a kind of varnish, which soon hardens, and forms a perfect defence from the rain. The varnish is so hard, and binds the eggs so firmly together, that, if the twig be carefully severed, the whole mass of eggs can be slipped off entire. As this varnish produces the same effect on eggs as lacquer does upon polished metal, preserving the surface and defending it from moisture, the insect is called the Lacquer, a word which has been corrupted into Lackey.

In wet weather the Lackey caterpillars prefer to remain in their silken home, leaving it only for the purpose of feeding. They never lose their way, because, like the larvæ of the little ermine moth, which has been already described, they continually spin a single silken thread as they go along, and are, therefore, provided with an infallible guide to the track. Before they change to the pupal state they leave the nest.

The larva of this species is a very prettily marked creature, the body being striped with blue and yellow and white. The moth itself is yellow, with a slight tinge of orange, and across the upper pair of wings runs a dark band edged on either side by a paler streak. As there is another allied species, which lives on various seaside plants, the present insect ought more properly to be called the Tree Lackey. The moth seems to be rather periodical and local; for, although specimens are found annually in most years, they swarm to such an extent in certain places, that whole rows of fruit trees are denuded of their leaves, and covered with the silken webs of the pretty but destructive caterpillars.

THE BROWN-TAILED MOTH is another of the arboreal insects, and spins a web very like that of the gold-tailed moth, which has already been described. In some seasons it is more numerous than in others, and occasionally seen in vast multitudes. This phenomenon is often observable among insects, as is well known to all practical entomologists, and in more than one instance the caterpillars of the Brown-tailed Moth have been so plentiful as to become a positive pest.

They are social larvæ, and, as they are hatched late in the autumn, they spin a joint web, in which they can be secure throughout the winter months. As the brood is mostly numerous, and as two or more broods may unite in forming a common dwelling, their habitation is extremely large, often enveloping several branches together with their twigs and leaves. Like the nest of the gold-tailed moth, it is divided into chambers, and is externally irregular in form, depending entirely for its shape upon the locality in which it is constructed.

Even in this country it is sometimes plentiful enough to annoy the farmer, who does not like to see his hedgerows disfigured by the silken tents spun by these caterpillars; but in

APOICA

France it has occurred in such hosts as to entail a serious loss upon the agriculturist, whole rows of trees having been stripped of their leaves, and the denuded branches covered with the sheets of web in which lay the destroying armies.

It is hardly possible to overrate the wonderful varieties of form that are assumed by the nests of insects,—varieties so bold and so startling that few would believe in the possibility of their existence without ocular demonstration. No rule seems to be observed in them ; at all events no rule has, as yet, been discovered by which their formation is guided ; neither has any conjecture been formed as to the reason for the remarkable forms which they assume.

Perhaps, of all the nests in the splendid collection of the British Museum, there are none that cause so much surprise as the wonderful group which is represented in this illustration. Many persons pass through the room, and even take some notice of the various nests with which they are surrounded, but they seldom notice the peculiarities of this group until pointed out to them. When, however, their attention is directed towards it, they never fail to express their surprise at so curious a structure, and their admiration of the manner in which these natural homes are constructed.

If the reader will refer to the illustration, he will see that the nests are by no means uniform in size or shape. The larger one, for example, which occupies the centre, rather exceeds ten inches in diameter, while the small nest at the end of the same branch is scarcely half as wide, and the others are of all the intermediate sizes. In shape, too, they differ, some being perfectly hexagonal, others partly so, while others again are nearly circular, though on a careful inspection they show faint traces of the hexagonal form.

We will now examine these nests, and see where they agree with and differ from each other.

In the first place, their upper surfaces are more or less convex, according to their size ; and whether they are circular or hexagonal, the convexity remains the same. This form is evidently intended for the purpose of making them weather-proof ; for the rain torrents that occasionally deluge the country would soon wash to pieces any nest whereon the falling drops could make a lodgment. The surface is therefore as smooth as that of the various pasteboard wasps which build in the forests of tropical America.

The upper surface being convex, it naturally follows that the under surface is concave, inasmuch as the cells are of tolerably equal length. In fact, the nests somewhat resemble very shallow basins with very thick sides, and bear an almost startling resemblance to the cap of a very large and very well-shaped mushroom, the central specimen being so fungus-like in form that, if it were laid on the ground in a waste and moist spot, it would soon be picked up as a veritable mushroom. The colour, too, is yellowish brown, and the surface has a kind of semi-polish that increases the resemblance.

In the nests of our common wasp, or hornet, the sheets of paper which form the exterior show plainly where each successive flake has been deposited, and the sweep of the insect's jaws is marked distinctly upon the yielding material. Even in the case of the few British species which build pensile nests in the open air, the separate flakes can be distinguished, though they are not so clearly marked as in those homes which are defended from the weather by earth or wood. Our temperate region knows no such sudden vicissitudes of weather as take place near the equator, and there is no need for insect habitations to possess very great strength or powers of resisting water. But in these nests the cover is so beautifully uniform, that no trace of a jaw can be detected upon it.

Agreeing in general appearance, the nests vary somewhat in colour. Of the eight specimens, the generality are of the mushroom-like hue which has already been mentioned. Others, however, rather vary in this respect, and the uniform yellowish brown is pleasingly diversified by patches of red. One of the nests, however, boldly departs from the general uniformity, the surface being not only reddish brown over its whole extent, but as rough as if made of sand-paper, or from the skin of a dogfish. One or two, again, are much darker than the others; while one is almost white, with only a tinge of grey.

Another point in these nests is, that although they vary so much in diameter, their thickness is almost uniform. The reason is evident enough. As the young larvæ attain a tolerably uniform size, and are not boldly divided into large males, larger queens, and little workers or neuters, the cells are of equal length. Therefore, whether the number be great or small, the thickness of the cell-group remains unchanged, though the diameter may increase to any reasonable amount.

All the nests are fixed in the same manner, a branch or twig passing through the upper surface. When the nest increases in size, the original support is often found to be too slight; and in that case, others are added. The smaller nests are upheld by a single twig only, but the largest is supported at no less than three points, two tolerably stout branches passing through the side of the cover, and a smaller twig supporting the top.

Another point to be noticed is, that the size of the nest is no criterion of its shape. It is not necessarily circular because it is large, nor hexagonal because it is small. The eight examples in the British Museum show every gradation of shape between the hexagon and the circle, without the least reference to size.

How the insect forms these wonderful cell-groups is an enigma to which not the least clue can be found. In proportion to the size of the architect, they are simply enormous, and yet the sides and angles are as true and just as if they were single cells. It is very clear that neither the theory of excavation or of equal pressure can apply to these nests, and an additional reason is afforded why these theories should be abandoned. It is to be regretted that the only reasoning is of the destructive kind; but at present we have no data on which to found a theory that seems in the least tenable.

In the nest to which reference has been made, the insects have carried out the hexagonal principle in a curious manner. A number of cells whose mouths are closed with a white silken cover prove that the inmates are undergoing their metamorphosis, and are in a transitional state between the larva and the perfect insect. Instead, however, of being scattered at random throughout the nest, the inhabited cells are arranged in the most systematic manner, a group occupying the centre, and being surrounded at a little distance with a row of covered cells which follow the shape of the exterior outline, and therefore take the shape of a hexagon.

The insect well deserves its scientific title. The generic name *Apoica* is formed from two Greek words, which signifies a colony, and the specific title *pallida* is given in reference to the hue of the body. It is not a handsome nor even a striking insect, being long, slender, and very pale yellow, looking as if it had once been decorated with a brighter covering. It has altogether a faded and semi-bleached look, suggesting to a practical entomologist that it had been subjected to sulphur-fumes, and thereby lost its colouring. Even the wings have the same pallid hue as the body, but with a white cast, and altogether the insect seems far too purposeless of aspect to construct houses which demand so much energy as those which we have just examined.

OUR last example of insect pensile nests is, I believe, one that has not yet been described, owing to its recent arrival in this country.

Whilst I was examining some specimens in the insect-room of the British Museum, two gentlemen brought for examination a box full of insect habitations, which they could not identify with those of any known species. At first sight they appeared to be specimens of galls, but a more careful inspection soon showed their real character. They were formed very much like those of the Housebuilder Moth (see page 166), but with a singular addition. Several specimens are now before me, which will be briefly described.

The foundation of the nest is a structure of leaf-stems and fragments of leaves, varying much in size, some being thicker than crowquills, and others as fine as ordinary needles. These are arranged cross-wise upon each other, so that the nest might easily be mistaken for that of a large caddis-worm. The nests, however, differ much in form, size, and material,—some being half as large again as others, and some being made almost entirely of large pieces of leaf, and others chiefly of stems, among which the leaf-fragments are closely pressed.

We will now proceed to cut open one of these nests in order to view its structure.

The outer covering is remarkably close, stiff, and tough, although very thin, and crackles like parchment as the scissors pass through it. When cut, it is found to be almost distinct from the nest which it covers, being only attached to the projecting ends of the leaf-stems, and so slightly fastened to them that it can be lifted off without injury, only leaving a few threads adherent to the stem.

We now turn back the severed flap, and the body of the nest comes to view. In the dry state the leaf-stems are so hard that they require a strong and sharp pair of scissors to penetrate them. I nearly broke a moderately fine pair of scissors in a vain endeavour to open the nest. Even in their fresh state the stems must have been tolerably strong, and the architect must have possessed a powerful pair of jaws for their severance. The stems are crossed upon each other, much as confectioners cross sticks of chocolate, so that the ends slightly protrude, and a hollow space is left in the centre. Pressed tightly among the sticks are fragments of leaves, not torn from the small delicate portions, but cut completely through the largest nervures, and seeming, indeed, as if the strongest parts of the leaves were intentionally selected. In the specimens now before me the upper surface of the leaf is always towards the exterior of the nest.

We now take a very strong and sharp pair of scissors, push one point into the nest, and carefully cut a flap corresponding with the severed portion of the silken cover. The flap is easily turned back, and discloses a smooth and silken lining, much resembling that which forms the cover. The lining, however, is softer than the cover, and does not crackle when bent. Thus we see that the nest consists of four distinct layers : first, the soft silken lining, then a cover of leaf-fragments, then a protecting *chevaux-de-frise* of stems, and lastly a cover of silk, so that the inhabitant is as well protected from weather and foes as can be imagined.

The next proceeding is to discover the architects of the nests. This is easily done, for some of the architects have assumed their perfect state during the voyage home, while others are preserved by spirits, in which their discoverer has thoughtfully placed some specimens.

Here I may be allowed to mention that the example set by Mr. W. J. Tomkinson, who sent over these interesting objects, is one which is well worthy of imitation. Residents in other countries are too apt to forget the interests of their own, and they soon become familiar with the objects which at first are new and strange to them, and at last become entirely indifferent. Even when they do take the trouble to collect and send home a few objects, they do so in such a manner that they are almost useless, no description being given of them, and no clue afforded which can help the home-staying student.

Here, however, proper pains have been taken, and the value of the objects is in consequence multiplied a hundred-fold. A number of nests were sent as they were collected from the branches, and, in order to show that the architect is not confined to one species of tree, they have been carefully selected from several trees, such as the oak, acacia, and alder. My specimens are taken from the last-mentioned tree. Knowing that the pupæ would become moths in the course of the voyage, Mr. Tomkinson placed a number of them in the box, so that a perfect series of the insect has been obtained, namely, the male and female, pupa and larva, some in the dried state and others in spirits, in order that the internal anatomy might be examined.

Before the male caterpillar changes into a chrysalis it reverses its position, so that the head is close to the orifice which was previously occupied by the tail. When it has completed its change, and is about to issue into the world, it forces itself out of the nest as far as the base of the abdomen. The female never leaves her home, and never changes her attitude, and scarcely changes her form. After she has emerged from the pupal states, she seems to return to her former condition, and would be taken by any ordinary observer for a caterpillar of more than ordinary fatness. She has no wings, and no legs to speak of, these members being needless in a creature that never changes her position. It is rather curious that the males should ever be able to find their spouses, but they are probably led by an instinct which we cannot comprehend, as is the case with several of the larger British moths.

The male is a rather small though stoutly made insect, and is not at all attractive in colour, being simple brown, with a few black markings on the wings. The antennæ, however, are very beautiful, being doubly feathered, like those of the Housebuilder Moth, to which the insect is closely allied, the feathering being widest at the base, and narrowing gradually to the tip. The whole of the body is clothed with long, dense, and soft hair, of a pale brown, and having a silken lustre. These beautiful nests were brought to the Museum by E. H. Armitage, Esq., who kindly presented me with the specimens which have been described.

A SOMEWHAT similar nest, but of a much more formidable aspect, was discovered by W. B. Lord, Esq., R.A., and has been figured in the *Boys' Own Magazine* for August, 1864. The shape of the nest is very remarkable, and is exactly that of a soda-water bottle, suspended by its neck. A very tolerable imitation of this curious nest could be made by coating a soda-water bottle with clay, and sticking it full of porcupine quills, with the points radiating on every side. The following is Mr. Lord's own description :—

'On looking closely at the thorny, sinuous branches, we shall see a number of little pendent prickly things, each hanging to its own silken cord, like juvenile hedgehogs "lynched" by the fairies of the spring.

'These are a peculiar species of "tree-caddis," which, as far as I know, are as yet undescribed by anyone. Their cases are curiously armed with thorns. nipped from the tree on which they hang. The thorns are all disposed with their points outwards, and are stuck into a strong, glutinous material of which the body of the case is composed, and they look for all the world like the spikes of *chevaux-de-frise*. A web-like skein of singularly strong material serves as a rope whereby to swing the caddis-case from the branch to which it is attached. And a nest more difficult to swallow, and hard to digest, its enemies would be rather puzzled to find.'

As is frequently the case with such nests, the peculiar form serves a double purpose, namely, protection and concealment, the sharp points of the thorns performing the former duty, and their similarity to surrounding objects the latter. Acacias are conspicuous for the thorns with which their branches and sometimes their trunks are studded, and in several species the wooden bayonets are several inches in length, and as large and sharp as porcupine quills. These thorns are crowded thickly on the branches, and always diverge from each other, so that the hand can scarcely be insinuated among the boughs without suffering several wounds. The nest being surrounded with these thorns, it is evident that all ordinary foes would be baffled by such an array of points, no matter how anxious they might be to get at the creature within.

The thorns are equally efficient as a means of concealment, for, as they are taken from the tree itself, they cause the nest to harmonise so perfectly with surrounding objects, that it is not very easily perceived.

As long as the caterpillar remains in its larval state, and is obliged to feed, it traverses the branches freely, carrying with it the prickly home, and bearing the whole of its weight as it moves. But when the pupal stage has nearly arrived, the nest is suspended to the branch by strong silken threads, and thenceforth remains immoveable.

CHAPTER XXXI.

MISCELLANEA.

The RAFT SPIDER—Why so called—Mode of obtaining prey—Mice and their homes—The CAMPAGNOL or Harvest Mouse—Its general habits—Its winter and summer nest—Its storehouse and provisions—Entrance to the nest—The WOOD MOUSE and its nest—Uses of the Field Mice—The DOMESTIC MOUSE—Various nests—Rapidity of nest-building—A nest in a bottle—The cell of the QUEEN TERMITE—Its entrances and exits—Size of the inmates—The CLOTHES MOTHS and their various species—Habitations of the Clothes Moth, and the method of formation and enlargement—The ELK and its winter home—The snow fortress and its leaguers—Its use, advantages, and dangers—The ALBATROS and its mode of nesting—Strange scenes—The EDIBLE SWALLOW—Its mode of nesting—Origin of its name—Description of the nest—The EAGLE and its mode of nesting—Difficulty of reaching the eyrie—The NIGHTINGALE and its nest—Other ground-building birds and their temporary homes—The NODDY—Perilous position of the eggs, and young—The COOT, and its semi-aquatic nest.

IN this, the concluding chapter, are described sundry habitations which cannot well be classed in any of the previously mentioned groups, and which present some peculiarities which render them worthy of a separate notice.

The reader will remember that the water spider is in the habit of constructing beneath the water a permanent home, to which it retires with the prey which it has caught, and in which it brings up its young. There is another spider which frequents the water, but which only makes a temporary and moveable residence. This is the RAFT SPIDER (*Dolomedes fimbriatus*) which is represented in the illustration of its natural size.

As may be seen by reference to the figure, it is a large species, being, indeed, one of the largest British spiders, its size depending more upon the dimensions of the body than the length of the limbs. It is a remarkably handsome spider, its general colour being chocolate-brown, and a broad orange band being drawn so as to mark the outline of the abdomen and thorax. There is a double row of small white spots upon the surface of the abdomen, and a number of short dark transverse bars give variety to the colouring. The limbs are pale red.

This creature belongs to that group of spiders which do not live in a web, and wait for casual insects, but which chase their prey after the manner of carnivorous vertebrates. Indeed, it

RAFT SPIDER.

may fairly be said to belong to the large group of wolf spiders, and is nearly allied to them.

The Raft Spider is only to be found in fenny or marshy places, and is mostly seen in the fens of Cambridgeshire, where its remarkable habits have long been known. Not content with chasing insects on land, it follows them in the water, on the surface of which it can run freely. It needs, however, a resting-place, and forms one by getting together a quantity of dry leaves and similar substances, which it gathers into a rough ball, and fastens with silken threads. On this ball the Spider sits, and allows itself to be blown about the water by the wind. Apparently, it has no means of directing its course, but suffers its raft to traverse the surface as the wind or current may carry it.

There is no lack of prey, for the aquatic insects are constantly coming up to breathe the air; and although they may only remain on the surface for a second or two, the Spider can seize them before they can gain the safe refuge of the deeper water. Then there are insects, such as the gnat, which attain their wings on the surface of the water, and can be taken by the Spider before they have gained strength for flight. Also, there are insects which habitually traverse the water in search of prey, and which are themselves seized by the more powerful and equally voracious Spider. More than this, moths, flies, beetles, and other insects, are continually falling into the water, and these afford the easiest prey to the Raft Spider, who pounces upon them as they vainly struggle to regain the air, and then carries them back to its raft, there to devour them in peace.

The Spider does not merely sit upon the raft, and there capture any prey that may happen to come within reach, but when it sees an insect upon the surface, it leaves the raft, runs swiftly over the water, secures its prey, and brings it back to the raft. It can even descend below the surface of the water, and will often crawl several inches in depth. This feat it does not perform by diving, as is the case with the water spider, but by means of the aquatic plants, down whose stems it crawls. Its capability of existing for some time beneath the surface of the water is often the means of saving its life; for, when it sees an enemy approaching, it quietly slips under the raft, and there lies in perfect security until the danger has passed away.

There is, living in the same localities, a closely-allied species, the PIRATE SPIDER (*Lycosa piratica*), which has similar habits, chasing its prey on the water, and descending as well below the surface. It does not, however, possess the power of making a raft.

In a previous chapter of this work, the beautiful pensile nest of the Harvest Mouse has been described and figured, and the burrows of other species of mouse have been cursorily mentioned. I shall now proceed to describe the nests of the common Field Mice, together with the habitation of the little brown-coated, long-tailed, sharp-nosed rodent, that is so familiar in houses unguarded by cats or traps.

We will first take the nest of the Short-tailed Field Mouse, otherwise termed Campagnol, or Field Vole (*Arvicola arvensis*). This pretty little creature, whose red back, grey belly, short ears, and blunt nose, might be seen daily if human eyes were more accustomed to observation, is extremely plentiful in the fields, especially those of a low-lying and marshy character, such as water meadows and hay-fields near rivers.

Though more nocturnal than diurnal in their habits, the little creatures are not afraid of daylight, and I have often captured them when the sun was at its meridian height. But they are so smooth and easy in their movements, harmonise so well with the colour of the soil, and glide so deftly between the grass, that they can scarcely be distinguished even when the blades are only a few inches in length. I have known them to traverse the ground while a game at cricket was proceeding, and to cross the closely-mown space between the wickets, as if serenely conscious of their invisibility.

They seem to glide rather than to walk, and thread their way silently and without noise. Even when the grass is short, a little patch of reddish earth attracts no attention, and the red-brown fur of the mouse is so similar to such earth, that few would notice it. But if a more attentive observer finds that in a few seconds the ruddy patch has changed its place, his suspicions are at once aroused, and he examines the moving tint more curiously. He must, however, keep his eye upon it as he moves towards it, for if he once loses sight of it, he will in all probability miss it altogether, and think that his eye must have deceived him.

Towards the evening, however, the Campagnol is less fearful, and not only traverses the fields, but ascends the shrubs and plants in search of food. It climbs nearly as well as a squirrel, its sharp nails hooking themselves into every irregularity of the bark, and its long finger-like toes clasping round the grass stems and little twigs like the claws of a monkey. An autumnal evening is the best time for watching the Campagnol, and if the observer will only remain perfectly quiet, and keep a good opera-glass in readiness, he will be greatly interested by the little animal. A hedge in which are plenty of dog-roses is a likely place for the Campagnol, as the animal is very fond of the ripe hips, and ascends the shrubs in search of its daily food. When it reaches the branch bending with the scarlet load, the mouse runs swiftly and sure-footed as a rope-dancer, and carries off a store of the fruit, partly for present consumption and partly for a stock of winter food.

For the little creature is not one of the hibernating animals, or, at all events, the semi-sleep is of so light a character that the mouse comes often abroad, even in the depth of winter. It is undeterred by severe frost, and takes little heed of snow, as is proved by its tiny footmarks being tracked in the white and yielding substance.

This little mouse makes two kinds of nest, one for the winter, and another for the summer. The winter nest is below ground, and is approached by a hole varying much in length. As the cavity in which the nest reposes is larger than the tunnel, and of a globular form, it is mostly usurped by the wasp when the Mouse deserts it for summer quarters. Sometimes it is placed at some depth in the ground ; but usually is only a few inches from the surface. This is the nest to which Burns refers in his well-known poem upon the Field Mouse whose nest he had inadvertently ploughed up.

Besides the winter nest itself, the animal has a storehouse or cellar in which are placed the provisions intended for winter use, when the weather prohibits the Mouse from leaving its home, or when the surrounding shrubs and bushes are plundered of their fruits and denuded of their bark. In this storehouse the animal conceals quantities of hips and other provisions, among which are found numbers of cherry-stones.

The summer nest is of entirely a different construction, being placed above ground, though tolerably well concealed. The following account of it, by Mr. J. J. Briggs, appeared originally in the *Field* newspaper. 'No wonder that in districts where they are difficult to keep down they increase with rapidity, for, like the common Mouse, they are prolific breeders. I have found nests of this Mouse in almost every week from the end of May to the middle of August, and each containing from one to ten young, usually from five to seven. The young look poor helpless creatures, being both blind and naked. They leave the nest in about a month, but remain with their parents for some time afterwards.

'The nest is placed on the ground in a pasture or meadow; a field of mowing grass is preferred, but I have found it among corn, where the long herbage affords the coveted quiet and concealment; but when the crop is cut, the nest is laid bare, and the young frequently fall a prey to hawks and other depredators. The nest is built in a little hollow on the surface of the earth, just concealed at the bottom of the stems of grass. If you pull it out it looks like a lump of herds or flax, being composed of numerous small pieces of grass nibbled to a fine texture with care by the parent animals.

'I have taken up dozens of nests to examine, but in no single instance could I ever find an entrance to the interior. How the parents gain admission to it seems extraordinary. This remark applies to the nest of the White-bellied Field Mouse, and White, of Selborne, notices the same fact with reference to the harvest-mouse. How the young are suckled seems marvellous, unless the conjecture be correct that the female opens a fresh aperture in the nest each time she visits her young, and closes it again when she departs.

'The parents show considerable affection for their young. If a nest be exposed by the mower they do not desert it, but on the contrary endeavour to conceal it from observation as well as they can, by drawing round it the neighbouring grasses and plants.'

The same writer remarks that he has several times caught the

Short-tailed Field Mouse in the hedges while 'bat-fowling' at night for small birds. He has also found that when the Mouse eats hips, it nibbles off one end and extracts the seeds, rejecting the husks as uneatable. Man, however, acts in just the reverse manner, rejecting the seeds with their cottony envelopes, and eating the sweet husk, or sometimes boiling it up with sugar and making it into a conserve.

The cherry-stones are mostly obtained through the agency of blackbirds, thrushes, and other feathered fruit lovers. These birds pluck the cherries, often leaving the stones adhering slightly to the stalks, or dropping them on the ground. In the former case the stones are sure to be flung down when the legitimate owner gathers the fruit, so that the Mouse who is fortunate enough to live in a cherry-growing district is sure of a winter stock of food. Several hundred cherry-stones are sometimes placed in a single storehouse, affording sustenance to several mice.

The animal eats them in a peculiar manner. Instead of splitting them open by using the chisel-edged teeth or wedges, after the manner of schoolboys opening nuts and peach-stones with their pocket-knives, the Mouse nibbles off one end of the stone so as to make a little hole, and through this small aperture it contrives to extract the solid kernel.

The LONG-TAILED FIELD MOUSE or WOOD MOUSE (*Mus sylvaticus*) also makes a winter nest, in which it lives, but to which it does not absolutely confine itself, making several nests in the course of a season, and selecting such spots as appear to please its fancy at the time. Mr. Briggs remarks that he has known one of these mice to make a nest in three days.

One species of Field Mouse sometimes does good service to mankind, through its habits of storing up its winter stock of provisions. Lately in the country about Odessa vast armies of mice were seen, and evidently did much damage. Not only did they eat the crops, but they swarmed into the houses in such numbers that traps could hardly be set fast enough, twenty or thirty being often taken in a single day.

Hurtful though they were in some senses, they nevertheless

had their uses. The country is liable to the attacks of locusts, which in that year happened to be particularly numerous. These destructive insects, as is the case with many of their order, lay their eggs enclosed in capsules, something like the well-known egg-cases of our too common cockroach. The mice were very fond of the egg-capsules, and not only devoured them as part of their daily food, but carried them away, laid them up in their treasuries for a winter store, thus thinning the locust armies far more effectually than man could have done.

We now come to the Common Mouse of our houses (*Mus musculus*).

This little animal is a notable house-builder, making nests out of various materials, and placing them in various situations. There seems to be hardly any place in which a Mouse will not establish itself, and scarcely any materials of which it will not make its nest. Hay, leaves, straw, bitten into suitable lengths, roots, and dried herbage, are the usual materials employed by this animal when it is in the country.

When it becomes a town mouse and lives in houses, it accommodates itself to circumstances, and is never in want of a situation for a nest or materials wherewithal to make a comfortable house. It will use up old rags, tow, bits of rejected cord, paper, and any such materials as can be found straggling about a house; and if it can find no fragments, it helps itself very unceremoniously, and cuts to pieces, books, newspapers, curtains, or garments.

Many instances of remarkable Mouse-nests are recorded, among which the following are worthy of mention.

As is usual, at the end of autumn, a number of flower-pots had been set aside in a shed, in waiting for the coming spring. Towards the middle of winter, the shed was cleared out, and the flower-pots removed. While carrying them out of the shed the owner was rather surprised to find a round hole in the mould, and therefore examined it more closely. In the hole was seen, not a plant, but the tail of a mouse, which leaped from the pot as soon as it was set down. Presently another mouse followed from the same aperture, showing that a nest lay beneath the soil. On removing the earth, a neat and comfortable nest was found, made chiefly of straw and paper, the entrance to which was the hole through which the inmates had fled.

The most curious point in connection with this nest was, that although the earth in the pot seemed to be intact except for the round hole, which might have been made by a stick, none was found within it. The ingenious little architects had been clever enough to scoop out the whole of the earth and to carry it away, so as to form a cavity for the reception of their nest. They did not completely empty the pot, as if knowing by instinct that their habitation would be betrayed. Accordingly, they allowed a slight covering of earth to remain upon their nest, and had laboriously carried out the whole of the mould through the little aperture which has been mentioned. The flower-pot was placed on a shelf in the shed, and the earth was quite hard, so that in the process of excavation there was little danger that it would fall upon the architects.

Another nest was discovered in rather an ingenious position. A bird had built a nest upon a shrub in a garden, and, as is usual in such cases, it had placed its home near the ground. A Mouse of original genius saw the nest, and perceived its value. Accordingly, she built her own nest immediately below that of the bird, so that she and her young were sheltered as by a roof. So closely had she fixed her habitation, that, as her young ran in and out of their home, their bodies pressed against the floor of the bird's nest above them. No less than six young were discovered in this ingenious nest.

Another very remarkable nest of the Common Mouse has been chronicled in the same journal to which reference has repeatedly been made. 'Early in March we set a hen; and, as her nest was a basket, a sack was placed under and around it, so as to keep in the heat. When the hen was set, she was in good feather, wearing an ample tail, according to her kind (the Brahma); but as the three weeks went on, her tail seemed much broken, assumed a dilapidated appearance, and finally became a mere stump. This excited notice and surprise, as there was nothing near her against which she was likely to spoil her tail.

'When the chickens were hatched, and they and their mother were taken to a fresh nest, and the old one removed, it was found that a Mouse had constructed a beautiful nest under the basket. The body of the nest was made of tow scraped from the sack, and chopped or gnawed hay from the hen's nest; while the lining was made of the feathers of her tail, which had evidently been removed, a small bit at a time, as wanted, until all the feathers were reduced to stumps, showing marks of the Mouse's teeth. We should have liked to have heard the hen's remarks on the transaction, when the Mouse was nibbling her tail.'

In this case the Mouse improved on the conduct of her relative that built in the garden; for, by placing her nest in such a position, she not only secured the very best materials for her home, but enjoyed the advantage of the regular and high temperature which proceeded from the body of the sitting hen, and which was admirably adapted for the well-being of her young family.

Our last example of a remarkable Mouse-nest is that which is figured in the accompanying illustration, and which was drawn from the actual object.

A number of empty bottles had been stowed away upon a shelf, and among them was found one which was tenanted by a Mouse. The little creature had considered that the bottle would afford a suitable home for her young, and had therefore conveyed into it a quantity of bedding, which she made into a nest. The bottle was filled with the nest, and the eccentric architect had taken the precaution to leave a round hole corresponding to the neck of the bottle. In this remarkable domicile the young were placed; and it is a fact worthy of notice, that no attempt had been made to shut out the light. Nothing would have been easier than to have formed the cavity at the underside, so that the soft materials of the nest would exclude the light; but the Mouse had simply formed a comfortable hollow for her young, and therein she had placed her offspring.

It is therefore evident that the Mouse has no fear of light, but that it only chooses darkness as a means of safety for its young.

The rapidity with which the Mouse can make a nest is somewhat surprising. One of the Cambridge journals mentioned, some few years ago, that in a farmer's house a loaf of newly-baked bread was placed upon a shelf, according to custom Next day, a hole was observed in the loaf; and when it was cut open, a Mouse and her nest were discovered within, the latter

MOUSE NEST IN BOTTLE.

having been made of paper. On examination, the material of the habitation was found to have been obtained from a copy-book, which had been torn into shreds, and arranged into the form of a nest.

Within this curious home were nine young mice, pink, transparent, and newly born. Thus, in the space of thirty-six hours at the most, the loaf must have cooled, the interior been excavated, the copy-book found and cut into suitable pieces, the nest made, and the young brought into the world. Surely it is no wonder that mice are so plentiful, or that their many enemies fail to exterminate them.

A GENERAL account of the TERMITES, or WHITE ANTS as they are popularly but erroneously called, has been given under the head of Building Insects, and it has been mentioned that the female, or queen, has a cell distinct from the habitation of her subjects, and that she never leaves it until her death. In order that the reader should understand more fully the structure of the royal cell, an illustration of it is here introduced.

When viewed from the outside, it would hardly be recognised for the habitation of an insect, for it looks like a large lump of hardened clay, about as large as an ordinary French roll, and not very unlike it in shape. On a closer inspection, a number of little holes may be seen, and these apertures afford an unfailing indication as to the real nature of the clay lump. Fig. 2 represents the external appearance of one of these cells.

Supposing that a queen Termite cell be cut vertically, so that the knife passes through either of the little round holes, it will present an appearance which is shown at Fig. 1. The large hollow of the cell is nearly filled by the body of the female, whose head and thorax are seen in the cavity. On either side is a section of the little holes, which are shown to be cylindrical passages communicating with the interior of the cell. The worker Termites, being very small, can traverse these passages with perfect ease, while the enormous body of the female is utterly unable to pass.

Through these passages the workers are continually passing, some entering with empty jaws, and others emerging, each holding between its mandibles an egg, which it is conveying to the nurseries. So rapidly are the eggs laid, that the workers are fully employed in carrying them out and placing them under the charge of the nurses.

The contrast in size between the workers and the queen can easily be seen by reference to the illustration. At Fig. 5 is shown the queen, and in the right hand of Fig. 1 is seen one of the workers passing through the tunnel. None but the workers can pass through so small an aperture, for the fighters or soldiers are of very much greater size than the workers, as may be seen at Fig. 4.

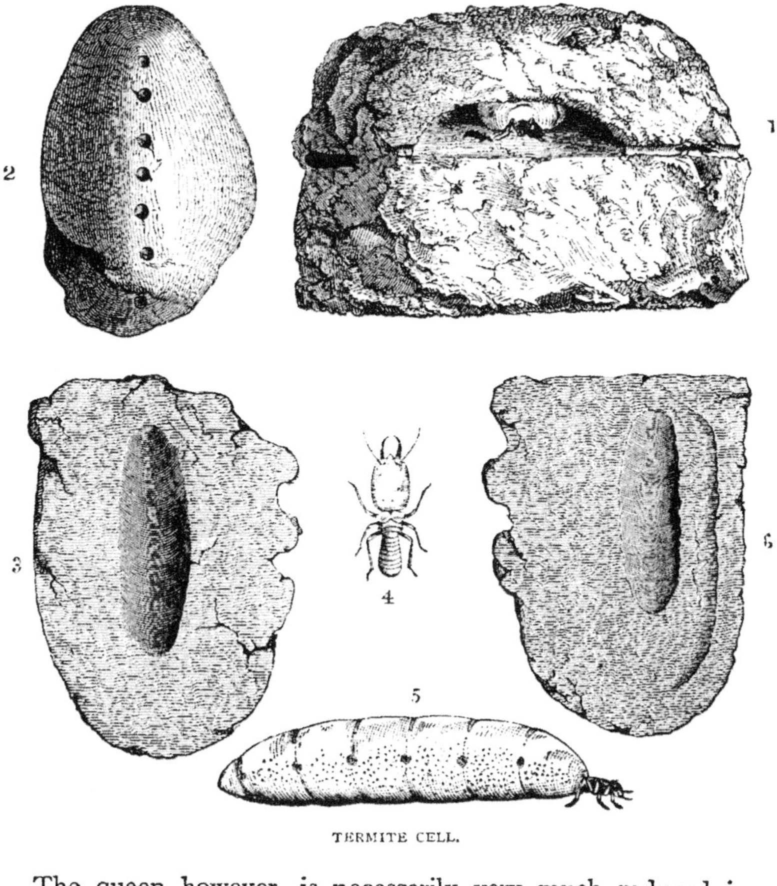

TERMITE CELL.

The queen, however, is necessarily very much reduced in size, as, if she had been drawn of her full dimensions, she would have occupied the whole length of the drawing. Before she is immured in the royal cell, she is by no means a large insect, the abdomen being in ordinary proportion to the thorax and head. But, when she has been fairly installed in her office her abdomen begins to enlarge, until it becomes so enormous that she is totally unable to move, and therefore, her enforced prison is so far from being a hardship, that it is a necessary protection for her huge and soft body, which is several hundred times larger than that of her mate. Large indeed she must be, as she is calculated to produce, on the average, rather more than thirty million eggs.

Figs. 3, 3 show the appearance of the royal cell when split open longitudinally, the recess which contains the queen being seen nearly in the centre. All the drawings are taken from specimens in the British Museum, and in the cell which is here figured, the outline of the queen is quite perceptible, having been impressed on the interior of the cell. The mode by which it is enlarged is also shown, a further enlargement having been begun, but cut short by the demolition of the nest. The cells vary very much in size, probably in accordance with the dimensions of the enclosed queen. I have seen them as large as cocoa-nuts, and of an extraordinary weight, the greater portion of the mass being solid clay.

THERE are many insects whose habitations are peculiarly annoying to mankind, and yet are extremely interesting to those who take an interest in the workings of instinct. Chief among these insects is the well-known CLOTHES MOTH. There are several allied species which popularly go by this name, but the most plentiful is that which bears the scientific title of *Tinea vestianella.* These destructive little creatures are proverbially injurious to clothes, especially if the garments be made of wool or furs, vegetable fabrics being not to their taste. Some species affect dried insects, and are in consequence extremely hateful to the entomologist; while their ravages on furs and feathers, and even on leather itself, render them the dread of those who, like myself, possess collections of natural history or ethnology.

In their winged state, the moths themselves do no direct

harm; but their young are doubly mischievous, firstly, because they devour the fabrics in which they live, and secondly, because they cut up the cloth, fur, or feathers, in order to obtain material for their home. Possibly for the sake of concealment as well as protection, the larva instinctively forms a habitation which entirely covers its white body, and which is almost imperceptible to the eye, because it is formed of the same materials as the fabric on which it lies.

The habitation is tubular in form, though not exactly cylindrical, being rather larger in the middle than at the ends, and open so as to allow the extremities of the caterpillar to protrude. One object in this structure is, to enable the inmate to turn in its cell, an operation which must necessarily be performed whenever the tubular home is enlarged. The process of enlargement is continually going on, and it is in consequence of this proceeding that so much material is used.

The manner in which the little creature enlarges its home is as follows:—

Without quitting its tubular home, it cuts a longitudinal slit throughout half its length or so, and opens the case to the required width. It then proceeds to weave a triangular piece of webbing, with which it fills up the opened slit, and rejoins the edges with perfect accuracy. As one end of the case is now larger than the other, the caterpillar turns its attention to the other end, cuts it open, widens it, and fills up the gap precisely as it had done to the first part. When the soft tube is sufficiently widened, it is lengthened by the addition of rings to each extremity.

By taking advantage of this peculiar method of house-making, observant persons have forced the Clothes Moths to make their tubular homes of any colour and almost of any pattern. By shifting the caterpillar from one coloured cloth to another, the required tints are produced, and the pattern is gained by watching the creature at work, and transferring it at the proper season. For example, a very pretty specimen can be produced by turning out of its original home a half-grown caterpillar, and putting it on a piece of bright green cloth. After it has made its tube, it can be shifted to a black cloth, and when it has cut the longitudinal slit, and has half filled it up, it can be transferred to a piece of scarlet cloth, so that the complementary colours of green and scarlet are brought into juxtaposition, and "thrown up' by the contrast with the black.

The caterpillar is not very particular as to the kind of material which it employs, and on which it feeds. Mr. Rennie makes the following observations on one of these creatures, whose proceedings he had watched. 'The caterpillar first took up its abode in a specimen of the ghost-moth (*Hepialus humuli*), where, finding few suitable materials for building, it had recourse to the cork of the drawer, with the chips of which it made a structure, almost as warm as it would have done from wool. Whether it took offence at our disturbing it one day, or whether it did not find sufficient food in the body of the ghost-moth, we know not; but it left its cork house, and travelled about eighteen inches, selected the "old lady" moth (*Mormo maura*), one of the largest insects in the drawer, and built a new apartment, composed partly of cork as before, and partly of bits clipped out of the moth's wings.

'We have seen these caterpillars form their habitations of every sort of insect, from a butterfly to a beetle, and the soft, feathery wings of moths answer their purpose very well; but when they fall in with such hard materials as the musk-beetle, or the large scolopendra of the West Indies, they find some difficulty in the building.

'When the structure is finished, the insect deems itself secure to feed on the materials of the cloth, or other animal matter within its reach, provided it is dry and free from fat or grease, which Réaumur found it would not touch. For building, it always selects the straightest and loosest pieces of wool; but for food it prefers the shortest and most compact; and to procure these, it eats into the body of the stuff, rejecting the pile or nap, which it necessarily cuts across at the origin and permits to fall, leaving it threadbare, as if it had been much worn.'

From the account which has just been given, it is evident that the caterpillar must be able to turn completely round in its case, and in order to enable it to perform this evolution, the tube is much wider in the middle than at the ends.

The instinct of the parent moth enables it to discover with astonishing certainty any substance which may afford food to its future young. Stuffed birds suffer terribly from the moth, because the arsenical soap with which the skins are preserved does not extend its poisonous influence to the feathers. I have known whole cases of birds to be destroyed by the moth, all the feathers being eaten, and nothing left but the bare skins.

Even the most deadly poison, corrosive sublimate, is not effectual, unless it settles on every feather. There is now before me a stuffed golden-eye duck, preserved by myself, the close plumage of which has partially thrown off the poisoned solution, and has consequently admitted the moth in small patches of feathers, especially about the neck. There is also in my collection a Kaffir shield, made of an ox-hide, which has been washed with the solution, and is almost entirely secure from the depredations of the moth. Yet there are one or two spots where a thong has protected the hair, and in those very spots the pertinacious moths have laid their eggs, and, in several instances, the caterpillars have succeeded in attaining their perfect state.

The ELK, or MOOSE (*Alces malchis*), inhabits the northern parts of America and Europe, and is, consequently, an animal which is formed to endure severe cold. Although a very large and powerful animal, measuring sometimes seven feet in height at the shoulders—a height which is very little less than that of an average elephant—it has many foes and is much persecuted both by man and beast. During the summer-time it is tolerably safe, but in the winter it is beset by many perils.

In its native country the snow falls so thickly, that the inhabitants of a more temperate climate can hardly imagine the result of a heavy storm. The face of the earth is wholly changed —well-known pits and declivities have vanished—white hills stand where was formerly a level plain—tier upon tier of mimic fortifications rise above each other, the walls being scarped and cut by the wind in weird resemblance of human architecture.

During the sharp frosts, the Elk runs but little risk, because it can traverse the hard, frozen surface of the snow with considerable speed, although with a strange, awkward gait. Its usual pace is a swinging trot; but so light is its action, and so long are its legs, that it quietly trots over obstacles which a horse could not easily leap, because the frozen surface of the snow, although competent to withstand the regular trotting force, could not endure the sudden impact of a horse when leaping. As an example of the curious trot of this animal, I may mention that on one occasion an Elk was seen to trot uninterruptedly over a number of fallen tree-trunks, some of which were nearly five feet in diameter.

It is a remarkable fact that the split hoofs of the Elk spread widely when the foot is placed on the ground, coming together again with a loud snap when it is raised. In consequence of this peculiarity, the Elk's progress is rather noisy, the crackling sounds of the hoofs following each other in quick succession.

Want of food is sometimes a danger to the Elk; but the animal is taught by instinct to clear away the snow, and to discover the lichens on which it chiefly lives. The carnivorous animals, however, are always fiercely hungry in the winter-time, and gain from necessity a factitious courage which they do not possess at other times. As long, however, as the frost lasts, the Elk cares little for such foes, as it can distance them if they chase it ever so fiercely, or oppose them if by chance it should find itself in a place where there is no retreat. They do not like to attack an animal whose skin is so thick and tough that, when tanned, it will resist an ordinary pistol-bullet, and which has besides, an awkward knack of striking with its fore-feet like a skilful boxer, knocking its foes over, and then pounding them with its hoofs until they are dead.

But when the milder weather begins to set in, the Moose is in constant danger. The warm sun falling on the snow produces a rather curious effect. The frozen surface only partially melts, and the water, mixing with the snow beneath, causes it to sink away from the icy surface, leaving a considerable space between them. The 'crust,' as the frozen surface is technically named, is quite strong enough to bear the weight of comparatively small animals, such as wolves, especially when they run swiftly over it; but it yields to the enormous weight of the Elk, which plunges to its belly at every step.

The wolves have now the Elk at an advantage. They can overtake it without the least difficulty; and if they can bring it to bay in the snow, its fate is sealed. They care little for the branching horns, but leap boldly at the throat of the hampered animal, whose terrible fore-feet are now powerless, and, by dint of numbers, soon worry it to death. Man, too, takes advantage of this state of the snow, equips himself with snow-shoes, and skims over the slight and brittle crust with perfect security. An Elk, therefore, whenever abroad in the snow, is liable to many dangers, and, in order to avoid them, it makes the curious habitation which is called the Elk-yard.

This winter home is very simple in construction, consisting of a large space of ground on which the snow is trampled down by continually treading it so as to form both a hard surface, on which the animal can walk, and a kind of fortress in which it can dwell securely. The whole of the space is not trodden down to one uniform level, but consists of a network of roads or passages through which the animal can pass at ease. So confident is the Elk in the security of the 'yard,' that it can scarcely ever be induced to leave its snowy fortification, and pass into the open ground.

This habit renders it quite secure from the attacks of wolves, which prowl about the outside of the yard, but dare not venture within; but, unfortunately for the Elk, the very means which preserve it from one danger only lead it into another. If the hunter can come upon one of these Elk-yards, he is sure of his quarry; for the animal will seldom leave the precincts of the snowy inclosure, and the rifle-ball soon lays low the helpless victims.

The Elk is not the only animal that makes these curious fortifications, for a herd of Wapiti deer will frequently unite in forming a common home.

One of these 'yards' has been known to measure between four or five miles in diameter, and to be a perfect network of paths sunk in the snow. So deep indeed is the snow when untrodden, that when the deer traverse the paths, their backs cannot be seen above the level of the white surface. Although of such giant size, the 'yard' is not by any means a conspicuous object, and at a distance of a quarter of a mile or so, a novice may look directly at the spot without perceiving the numerous paths. This curious fact can easily be understood by those of my readers who have visited one of our modern fortifications, and have seen the slopes of turf apparently unbroken, although filled with deep trenches.

There are many other animals which form temporary habitations in which they can remain concealed, because they are taught by instinct how to make their domicile harmonise with the surrounding objects.

One very familiar instance may be found in the common Hare, whose 'form' is large enough to shelter the owner, and yet is so inconspicuous that the animal often lies undiscovered, though a human being has passed within a couple of paces of its home. The Hare is never at a loss for a home, and will often hide itself very effectually in a tuft of grass that seems scarcely large enough to conceal a rat. But it is by no means insensible of the value of a denser cover, and seems to have a peculiar affection for a thick, though small, clump of furze.

Within a mile or two of my house there is a heath which is partly studded with furze bushes, and which is a very paradise for various field animals. The field mice have covered it with their 'runs,' which are often so slightly below the surface, that if the finger be inserted in the entrance it can be pushed along the whole length of the burrow, the only cover being a slight layer of still living moss. As to the Hares, a 'form' can be found every few yards, and if a little thick stubbly furze-bush should be seen standing alone, it is nearly certain to be the home of a Hare, which has made its warm soft couch within the mass of needle-like prickles.

The TIGER has a very similar habit, and takes advantage of a certain drooping shrub, called the *Korinda*, which is of low growth, making its lair underneath the boughs, which afford at once a shelter from the sun and a concealment from enemies.

WE now pass to the Birds, the first of which is that remarkable species called the EDIBLE or ESCULENT SWALLOW (*Collocalia nidifica*). The popular name is given to it, not because itself is edible, but because its nest is eaten in some countries.

We have all heard of birds'-nest soup, and some of us may possibly have imagined that the nests in question are made of the ordinary vegetable substances, such as moss, leaves, and twigs. Some persons have thought that the material is fish spawn, while others think that it is secreted by certain glands situated in the throat, and therefore produced entirely by the bird. The real material is clearly a kind of seaweed. I possess some of this substance, which, when dried, is colourless and translucent, exactly like the nest. When placed in boiling water, it swells into a gelatinous mass, quite tasteless, as is the nest itself, and capable of being drawn into fibres like those of which the nest is made.

When first made, these nests are very white and delicate in their aspect, and in that condition are extremely valuable, being sold at an extravagant price to the Chinese. They soon darken by use and exposure, and are not fit for the purposes of the table until they have been cleaned and bleached.

These nests are found in Borneo, Java, &c., and are extremely local, being confined to certain spots. The birds always choose the sides of deep cavernous precipices, so that the task of obtaining the nests is extremely dangerous. They are attached to the perpendicular rocks much as the ordinary mud-built swallow-nests, and are generally arranged in horizontal layers. The caverns in which the nests are placed are extremely valuable, and are preserved with jealous care from any intruder.

One of these nests in my own collection is shaped much like one of the halves of a bivalve shell, and is thick at the base where it was attached to the rock, diminishing towards the extremity. On the outside it has a very shelly appearance, being made in regular layers, whose edges are as distinct as those of the oyster-shell, but which have a double and not a single curve. In shape it is somewhat oval, but the base is necessarily flat, on account of its attachment to the rock.

EDIBLE SWALLOW.

The material is so translucent, that when placed on printed paper and held to the light, the capital letters can be plainly read through its substance. A glance at the interior shows at once the mode of its construction. It is made of innumerable glutinous threads, which have been drawn irregularly across each other, and have hardened by exposure to the air into a material which much resembles isinglass. The natives say that the construction of a single nest occupies a pair of birds for two full months; so that there is some probability that the material may really be secreted by the birds themselves.

The nests are only used for one purpose. They are steeped in hot water for a considerable time, when they soften into a gelatinous mass, which forms the basis of a fashionable soup, and is not unlike the green fat of the ordinary turtle. Indeed, those who have partaken of birds'-nest soup say, that if it were seasoned in a similar manner, it might easily be taken for turtle soup. The Chinese value this soup highly, thinking that it possesses great power of restoring lost strength. It is, however, far too costly to be obtained by any but the rich, the best quality fetching rather more than sixty shillings per pound.

There are at least four species of swallow that make these curious nests, and the natives say that the entrance to the caves is always occupied by another kind of swallow, which makes a nest of mixed moss and gelatine, and which fights the valuable birds and drives them away. They therefore always attack the intruders, and endeavour to knock down their nests with stones. The nests are very small and shallow, and seem scarcely capable of accommodating either eggs or young birds. My own specimen is exactly two inches in length, one inch and three-quarters in breadth at its widest point, and scarcely more than half-an-inch deep. Its internal shape is exactly that of a spoon-bowl, one-third of which has been cut off abruptly near the handle.

NONE of the purely predacious birds are remarkable for their skill in architecture, and the EAGLE (*Aquila chrysaëtos*) is no exception to the general rule. The nest of this magnificent bird is nothing more than a huge mass of sticks flung at random on some rocky ledge, and having a shallow depression in which the young can lie. In general shape, or rather in shapelessness, it is not unlike the nest of the osprey, which has already been described, and it is so rudely put together that the sticks seem to afford even a less commodious bed than the bare rock.

The portion that is occupied by the young is comparatively

EAGLE.

small, and the general platform of the nest serves as a sort of larder, on which are deposited the birds, hares, lambs, and other animals which the parents have killed and brought home. Sometimes the nest will be amply supplied with food, but sometimes the parent birds are obliged to hunt daily. Young eagles are voracious beings, and if there be no sheep flocks within reach, the task of supplying them with food is a very heavy one, especially when they have nearly attained maturity. In feeding its young for the first few weeks of their life, the eagle tears the prey into little pieces, and impartially distributes the bleeding morsels to the gaping and screaming offspring. Afterwards, however, when the young eagles have gained strength of beak, the prey is merely dropped near them, and they tear it to pieces for themselves.

Generally the nest of the Eagle is placed in some inaccessible spot, and the bird seems never to be so pleased as when it can find a rocky ledge situated about half-way down a precipice, and sheltered from above by a large projecting piece of rock. This projection answers two purposes. It prevents the nest from being seen from above, and also guards it from being harried by persons let down by ropes. To take an Eagle's-nest is always a task of extreme difficulty, and one which tries to the utmost the nerves and endurance of the climber. It also makes considerable demands on his courage, for if the parent birds should discover the intruder, they are sure to attack him, and may very probably dash him to the ground.

Should the bold cragsman succeed in reaching the nest, he does not find it a very pleasant locality. The nostrils of the Eagle are very useful for the purpose of respiration, but the bird has apparently little or no olfactory sensibilities. The stench that arises from an inhabited Eagle's-nest is quite beyond the power of description, for the young Eagles themselves are not the sweetest beings in the world, and their evil odour is supplemented by that which arises from the refuse food that is suffered to putrefy in the very nest.

THERE are very many sea-birds which hatch their young on the shelves of precipitous rocks, and of them I have chosen for an example the bird which is called the NODDY (*Anous stolidus*). It is a species of Tern, and has long been celebrated among sailors for the ease with which it can be captured, especially if the daylight has departed.

The Noddy mostly chooses for its nesting-place some lofty precipice, and generally lays its eggs upon a shelf of the rock. Sometimes, but rarely, it takes a fancy to some low and thick bush, and in any case is but an indifferent architect. Often the nest is nothing more than a heap of seaweed, on the top of which is excavated a very slight hollow; and in no case does the bird seem to exercise any skill in the disposition of materials. As it returns year after year to the same spot, and never clears away the old nest, it manages in time to accumulate a heap of seaweed that is sometimes more than two feet in thickness, and of considerable width. The bird is gregarious in its nesting, the rocky ledges being crowded with the rude nests, and the odour that proceeds from them being absolutely intolerable to human nostrils. The eggs are rather pretty, being of an orange colour, spotted and splashed with red and purple of different shades.

It is rare in England, but there are many British birds that build in a similar manner, such as the Solan goose, or gannet, the cormorant, the guillemot, and various gulls.

THE nest of the NIGHTINGALE (*Luscinia Philomela*) could hardly be classed in any of the preceding groups, and therefore takes its place among the miscellaneous habitations.

It is not built in the branches, nor in a hole, nor suspended from a bough, nor absolutely on the ground. It is always set very near the ground, and in most cases it is scarcely raised more than a few inches above the soil. In one sense it is not a pretty nest. It is certainly not a neat one, and its apparent roughness of construction is probably intended to make it less conspicuous. The discovery of a Nightingale's-nest is not an easy task, unless the eye be directed to the spot by watching the movements of the bird. It is always most carefully hidden under growing foliage, and so well is it concealed, that even in places where nightingales abound, the detection of a nest is always welcome to the egg-hunter.

The materials of the nest are equally calculated for concealment, consisting of straw, grass, little sticks, and dried leaves, all being jumbled together with such 'artless art,' that even when a nest is seen, its real nature often escapes the discoverer. If the same materials were seen in a branch at any height from the ground they would at once attract attention, but in the position which they occupy they look like a mass of loose débris that has been blown by the wind and arrested by the foliage among which it has lodged.

The eggs are equally inconspicuous, being dull olive brown, without a spot or streak. After they are laid, the lively song of the Nightingale becomes less and less frequent, while after the young are hatched, the bird is silent until the next season. The Nightingale is as anxious to conceal itself as its nest, and seldom intentionally shows its brown plumage, though it will sing within six feet of a listener who will remain quiet. In the spring the bird seems as if it *must* sing, no matter who may be near, and its spirit of rivalry is so great, that the 'jug-jug' of one nightingale is sure to set singing all the others within hearing.

The Wandering Albatros (*Diomedea exulans*), the giant of the petrel tribe, makes its nest after a peculiar fashion.

It chooses the summit of lofty precipices near the sea, and its nest may be found most plentifully in Tristan d'Acunha and the Marion Islands. The Albatros is lord of the country, and no other living being seems to intrude upon its nesting place. So completely do the birds feel themselves masters of the situation, that if a human being penetrates to their haunts, they quietly move about as if he were non-existent, and do not appear to take the least notice of him. On such elevated positions the cold is necessarily intense, but the Albatros cares not for the cold, and brings up its white-coated young in a temperature that few human beings like to endure longer than needful.

No particular bed seems necessary for the egg, for the mother bird simply deposits it on the bare ground, and then scrapes earth round it so as to form a small circular wall. If their nest be approached very closely, the alarmed parents snap their bills like angry owls, and if they wish to be very aggressive they discharge from their bills a quantity of oil; but they seem to have no ideas of actual fight. The Albatros lays only one egg.

OUR last sample of 'Homes without Hands' is the ingenious structure that is made by the COOT (*Fulica atra*), the BALD COOT as it is sometimes called, on account of the horny plate on the forehead, which is pink during the breeding season, and white during the rest of the year. Although the general colour of the Coot is black, it is a pretty bird when in the water, and if the day be calm, the reflection on the surface has a very curious effect, the white patch appearing as if it rose to the surface of the water every time that the bird nods its head in the act of swimming.

The favourite nesting places of the Coot are little islands on which the grass grows rankly. Failing them it will make its nest among reeds and rushes, binding and twisting them together until they are firm enough to support the weight of the nest, the bird, and the many eggs. Should it not find either of these localities, it will build on the edge of the water, and almost invariably contrives to make its nest in such a manner that it cannot be reached from the land. The quantity of reeds, bulrushes, sedges, grass, and other materials used in the nest is very surprising; and yet, in spite of its large dimensions, it is not a conspicuous object. The nest contains a great number of eggs, seldom less than seven, and sometimes twelve or fourteen. They are whitish, and profusely spotted with irregular brown marks.

INDEX.

A

Aard Vark, 25
Accipiter, 290
Acheta, 98
Ægeria, 114
Agelena, 178
Agricultural Ant, 221
Albatros, 405
Alcedo, 35
Alces, 395
Amazon Ant, 282
Amphitrite, 243
Andrena, 81
Anous, 403
Ant, Agricultural, 221
— Amazon, 282
— Brown, 78
— Coushie, 74
— Driver, 274
— Dusky, 77
— Saüba, 74
— White, 207, 390
Ants, Foraging, 217
Ant-eater, Porcupine, 30
Ant-lion, 99
Anthidium, 110
Anthrocera, 169
Apoica, 370
Aquila, 401
Arctia, 168
Ardea, 329
Argyronetra, 229
Anomma, 274
Arvicola, 382
Athene, 16
Atlas Moth, 165
Atta, 221
Atypus, 67

B

Badger, 13
Bajjerkeit, 25
Baltimore Oriole, 142
Batikin, 289
Baya Sparrow, 148
Bear, Arctic, 11
— Black, 22
— Brown, 22
— Polar, 21
— White, 21
— Woolly, 168
Beaver, 245
Bee, Hive, 259
— Humble, 82
— Hoop-shaver, 110
— Lapidary, 84
— Rose-cutter, 107
— Willow, 107
Beetle, Burying, 92
— Clock, 94
— Dor, 94
— Musk, 106
— Sexton, 92
— Tiger, 91
— Wasp, 105
— Watchman, 94
Bettong, Brush-tailed, 180
— Pencilled, 180
Bettongia, 180
Bird, Bower, 202
— Cataract, 130
— Cow, 288
— Fire, 143
— Hanging, 144
— Lyre, 201
— Oven, 184
— Tailor, 124
Birgus, 50
Blackbird, 188
Blue-faced Honey-eater, 289
Bob-o'-link, 145
Bombus, 82-84
Boring Snail, 53
Bosjesman, 1
Bower-bird, 202
Brazilian Wood-nymph, 141
Breeze-fly, 316
Breeze-fly Clear-wing, 114
Brown Ant, 78
Brown-tail Moth, 272-367
Brush-tailed Bettong, 180
Bullen-Bullen, 201
Bullfinch, 334
Burnet Ichneumon, 161
Burnet Moth, 169
Burrowing-owl, 16
Burying Beetle, 92

C

Caddis, 233
— Shrimp, 244
— Tree, 377
Campagnol, 382
Campanular Wasp, 154
Canada Pouched-Rat, 19
Capocier, 347
Cassicus, 146
Cassique, Crested, 146
Castor, 245
Cataract Bird, 130
Celery Fly, 176
Cerambyx, 106
Cerapus, 244
Chaffinch, 330
Chat, Yellow-Breasted, 337
Chelidon, 190
Chelura, 51
Chicken, Mother Carey's, 41
Chigoe, 314
Chilabothrus, 45
Chipmuck, 18
Chipping Squirrel, 18
Chlamyphorus, 23
Chrysididæ, 296

Churr-worm, 97
Cicindela, 91
Clear-wing Moths, 114
Cleonus, 313
Clerus, 318
Clisiocampa, 367
Clock Beetle, 94
Clothes Moth, 392
Clytus, 105
Cnethocampa, 366
Coccygus, 288
Cojera, 30
Collocalia, 399
Columba, 39, 339
Coot, 406
Corvus, 39, 326
Cossus, 112
Cotile, 32
Coushie Ant, 74
Cow-Bird, 288
Crab, Fighting, 49
— Land, 47
— Racer, 49
— Robber, 50
Crested Cassique, 146
Cricket, Mole, 96
— Field, 98
Croaker, 97
Crow, 326
Cryptus, 294
Cuckoo, 288
Cuckoo Fly, 297
Curruca, 345
Cushat, 339
Cynipidæ, 297

D

Date, Shell, 59
Dicæum, 133
Digger, 1
Diomedea, 405
Dog, Prairie, 14
Dolomedes, 379
Dor Beetle, 94
Dormouse, 320
Dove, Ring, 339
— Rock, 339
— Stock, 339
— Turtle, 341
Drilus, 318
Driver Ant, 274
Drymoica, 347
Dusky Ant, 77
Duck-bill, 27
Dutchman's Pipe, 152

E

Eagle, 401
Earth-fly, 314
Echidna, 30
Eciton, 217
Edible Swallow, 399
Elephant-shrew, 10
Elk, 395
Emerald Throat, 353
Enneoctonus, 356
Entomophila, 132, 133
Entomyza, 289
Epeira, 66, 177
Ephemera, 98
Eriogaster, 170
Esculent Swallow, 399

F

Fairy Martin, 189
Fan-tailed Warbler, 125
Fiber, 182
Field Cricket, 98
— Mouse, Long-tailed, 385
— Short-tailed, 382
— Vole, 382
Fiery Topaz, 350
Fighting Crab, 49
Fire-bird, 143
Flies, Gall, 297
Fly, Breeze, 316
— Celery, 176
— Cuckoo, 297
Flycatcher, Great Crested, 147
— Red-eyed, 147
— White-eyed, 148
Fly, Earth, 314
Fly, Ruby-tailed, 296
Foraging Ants, 217
Formica, 77, 78
Fox, 11
Fratercula, 37
Fringilla, 330, 333
Fulica, 406
Furnarius, 184

G

Gall Flies, 297
Gall-Fly, Thistle, 314
Galleria, 116
Gallinula, 341
Gaper, 56
Garden Spider, 177
Gasterosteus, 225, 227
Gastropacha, 170
Gecarcinus, 47
Gelasimus, 49
Geometric Spider, 177
Geotrupes, 94
Goat Moth, 112
Gold-capped Weaver-bird, 123
Golden Oriole, 335
Golden Robin, 143
Goldfinch, 333
Gold-tailed Moth, 271
Goliathus, 96
Gopher, 19
Grakle, Purple, 291
Grallina, 186
Great-crested Flycatcher, 147
Great Grey Shrike, 356
Grey-throated Hermit, 139
Gribble, 52
Grosbeak, Sociable, 251
Gryllotalpa, 96
Grypus, 141

H

Hackee, 18
Hanging Bird, 143
Hang-nest, 143
Hare, 398
Harvest Mouse, 117
Hawk, Sparrow, 290
Helix, 54
Hen, Moor, 341
— Water, 341
Hermit, 352
— Grey-throated, 139
— Little, 137
— Pigmy, 139
Heron, 328
Hill-star, White-sided, 140
Hirundo, 189-191
Hive Bee, 259
Honey-comb Moth, 116
Honey-Eater, Blue-faced, 289
— Lanceolate, 135
— Painted, 132
— Singing, 130
— White-throated, 133
Hoop-shaver Bee, 110
Hornbills, 193

Hornet, 267
House-builder Moth, 166
House Martin, 190
Humble Bee, 82
Humming-Bird, Sawbill, 141

I

Icaria, 362
Ichneumon, 293
Ichneumon, Burnet, 161
Icteria, 337

J

Jackdaw, 39
Jannocumbine, 30
Jerboa Kangaroo, 180
Jigger, 314

K

Kangaroo, Jerboa, 180
Kestrel, 291
Kingfisher, 35
Korwé, 193

L

Lackey Moth, 367
Lamprosa,, 294
Lanceolate Honey-eater, 135
Land Crab, 47
Lanius, 356
Lapidary Bee, 84
Lazotænia, 173
Leaf-miners, 175
— rollers, 171
Lepus, 17
Lictor Moth, 167
Limnoria, 52
Limpet, 57
Lion, Ant, 99
Lithodomus, 60
Little-egger Moth, 170
Little Hermit, 137
Long-tailed Field Mouse, 385
— Titmouse, 195
Luscinia, 404
Lycosa, 67, 381
Lyre Bird, 201

M

Macroscelides, 10
Magpie, 199
Mahali Weaver-bird, 122
Mallangong, 28
Manis, 25
Martin, Fairy, 189
— House, 190
— Sand, 32
Mayfly, 98
Megachile, 107
Meles, 13
Menura, 201
Microgaster, 161, 293
Micromys, 117
Mischocyttarus, 365
Mole, 2
— Cricket, 96
— Shrew, 10
— Water, 27
Moor Hen, 341
Moose, 395
Moth, Atlas, 165
— Brown-tailed, 272
— Brown-tail, 367
— Burnet, 169
— Clothes, 392
— Goat, 112
— Gold-tailed, 271
— Honeycomb, 116
— Housebuilder, 166
— Lackey, 367
— Lictor, 167
— Little-egger, 170
— Oak-egger, 170
— Processionary, 366
— Small-ermine, 270
— Tiger, 168
— Vapourer, 168
— Wolf, 115
Mother Carey's Chicken, 41
Mouse, 386
— Harvest, 117
Mulo, 19
Mud Wasp, 285
Mus, 385, 386
Muscardinus, 320
Muscicapa, 147, 148
Musk Beetle, 106
— Rat, 10, 182
Musquash, 182
Mya, 56
Myrapetra, 156
Myrmeleon, 99
Myrmica, 273, 284
Myogalea, 10

N

Nectarinia, 159
Nicobejan, 30
Nightingale, 404
Noddy, 403
Norwegian Wasp, 153
Northern Wasp, 154

O

Oak-egger Moth, 170
Oak Moth, 171
Ocypode, 49
Œcodoma, 74
Œstrus, 316
Oiketicus, 165, 166
Ondatra, 182
Orchard Oriole, 145
Oreotrochilus, 140
Origma, 130
Orgyia, 168
Oriole, Baltimore, 142
— Golden, 335
— Orchard, 145
Ornithorhynchus, 27
Orthotomus, 124
Orycteropus, 25
Osmia, 108
Oven-bird, 184
Owl, Burrowing, 16

P

Painted Honey-eater, 132
Paniscus, 294
Parus, 195
Pelopæus, 214
Pencilled Bettong, 180
Pensile Spiders, 360
Petrel, Stormy, 39
Phaëthornis, 137, 139, 352
Philetærus, 251
Pholas, 58
Piddock, 58
Pica, 199
Pichiciago, 23
Pied Grallina, 186
Pigeon, Wood, 339
Pigmy Hermit, 139
Pimpla, 294
Pipe, Dutchman's, 152
Pirate Spider, 381
Platypus, 27
Plectorhynchus, 135

Plioposser, 122
Ploceus, 122, 123
Polar Bear, 21
Polistes, 158, 163
Politician, 148
Polybia, 256
Polyergus, 282
Porcupine Ant-eater, 30
Porthesia, 271, 367,
Pouched Rat, Canada, 19
Prairie Dog, 14
Processionary Moth, 366
Pseudostoma, 19
Ptilinus, 104
Ptilonorhynchus, 202
Ptilotus, 130
Puffin, 37
Pulex, 314
Purple Grakle, 291
Pyrrhula, 334

Q

Quiscalus, 291

R

Rabbit, 17
Racer Crab, 49
Raft Spider, 379
Raphigaster, 365
Rat, Canada, Pouched, 19
— Musk, 10, 182
Rattlesnake, 16
Razor Shell, 60
Red-backed Shriker, 356
Red-eyed Flycatcher, 147
Reed Warbler, 345
Ring-dove, 339
Robber Crab, 50
Rock-dove, 339
— Warbler, 130
Rook, 323
Rose-cutter Bee, 107
Ruby Throat, 352
— tailed Fly, 296

S

Sabella, 241
Saturnia, 165
Scarabæus, 94
Salicaria, 125, 344
Sand Martin, 32
Sauba Ant, 74
Saw-bill Humming Bird, 141
Scolia. 81
Scolytus, 102
Scorpion, 64
Sedge Warbler, 344
Sericornis, 127
Sexton Beetle, 92
Sheldrake, 39
Shell-binder, 237
Shipworm, 61
Short-tailed Field-mouse, 382
Shrew-elephant, 10
— mole, 10
Shrike, Great Grey, 356
—Red-backed, 356
Shrimp, Caddis, 244
— Wood-boring, 51
Singing Honey-eater, 130
Sirex, 110
Small Ermine Moth, 270
Snail, Boring, 53
Snake, Yellow, 45
Sociable Grosbeak, 251
— Weaver Bird, 251
Solen, 60
Sparrow, Baya, 148
— Hawk, 290
Spermophilus, 14
Spider, Garden, 177
— Geometric, 177
— Pensile, 360
— Pirate, 381
— Raft, 379
— Trap-door, 68
— Water, 229
Squirrel, 119
— Chipping, 18
Starling, 291
Sticklebacks, 225, 227
Stockdove, 39, 339
Stormy Petrel, 39
Sturnus, 291
Swallow, Chimney, 191
— Edible, 399
— Esculent, 399

T

Tachina, 297
Tadorna, 39
Tailor Bird, 124
Tambreet, 28
Tamias, 18
Tarantula, 67
Tatua, 151
Tauóca, 217
Teredo, 61
Termite, 207, 390
Tephritis, 176
Terebella, 237
Testudo, 44
Thalarctos, 21
Thalassidroma, 39
Thalurania, 142
Thistle Gall-Fly, 314
Thrush, 188
Tiger, 399,
— Beetle, 91
— Moth, 168
Tinea, 115, 392
Tinnunculus, 291 314
Titmouse, Long-tailed, 196
Tockus, 193
Tohunbuck, 28
Topaz, Fiery, 350
Topaza, 350
Tortoise, 44
Tortrix, 171
Toucan, 193
Trap-door Spider, 68
Tree Caddis, 377
Trichoptera, 233
Trochilus, 352
Troglodytes, 200
Trypoxylon, 214
Turdus, 188
Turtle Dove, 341

U

Urophora, 314

V

Vapourer Moth, 168
Vespa, 85, 153, 154, 215, 267
Vulpes, 11
Vole, Field, 382

W

Warbler, Fan-tailed, 125
— Reed, 345
— Rock, 130
— Sedge, 344
Wasp, 85, 215, 267

Wasp Beetle, 105
— Campanular, 154
— Mud, 285
— Northern, 154
— Norwegian, 153
Watchman Beetle, 94
Water Hen, 341
— Mole, 27
— Spider, 229
Weasels, 13
Weaver Bird, Gold-capped, 123
— Mahali, 122
— Sociable, 251
Whip-Tom-Kelly, 147
White Ant, 207, 390
White Bear, 21
White-eyed Flycatcher, 148
White-sided Hill Star, 140
White-throated Honey-eater, 133
Willow Bee, 107
Wish-ton-wish, 15
Wolf Moth, 115
Woolly Bear, 168
Wood-boring Shrimp, 51
Woodnymph, Brazilian, 141
Woodpeckers, 42
Wood-pigeon, 339
Woodquest, 339
Wren, 200
Wurbles, 317

X

Xanthornis, 145

Y

Yellow-breasted Chat, 337
Yellow Snake, 45
Yphantes, 142
Yponomeuta, 270

LONDON: PRINTED BY
SPOTTISWOODE AND CO, NEW-STREET SQUARE
AND PARLIAMENT STREET

Printed in Dunstable, United Kingdom

91219917R00245